Principles and Standards for School Mathematics Navigations Series

NAVIGATING
through
MEASUREMENT
in
GRADES 9–12

Masha R. Albrecht
Maurice J. Burke
Wade Ellis, Jr.
Dan Kennedy
Evan M. Maletsky

Maurice J. Burke
Grades 9–12 Editor

Peggy A. House
Navigations Series Editor

NATIONAL COUNCIL OF
TEACHERS OF MATHEMATICS

MAISD Regional
Math & Science Center
1001 Wesley Avenue
Muskegon, MI 49442

Copyright © 2005 by
The National Council of Teachers of Mathematics, Inc.
1906 Association Drive, Reston, VA 20191-1502
(703) 620-9840; (800) 235-7566; www.nctm.org

All rights reserved

Library of Congress Cataloging-in-Publication Data

Navigating through measurement in grades 9–12 / Masha Albrecht ... [et al.]; Maurice Burke, grades 9–12 editor.
  p. cm. — (Principles and standards for school mathematics navigations series)
  Includes bibliographical references.
  ISBN 0-87353-546-4
  1. Mensuration—Study and teaching (Higher) I. Albrecht, Masha. II Burke, Maurice Joseph. III. Series.
  QA465.N378 2004
  510'.71'2—dc22
                            2004020077

The National Council of Teachers of Mathematics is a public voice of mathematics education, providing vision, leadership, and professional development to support teachers in ensuring mathematics learning of the highest quality for all students.

Permission to photocopy limited material from *Navigating through Measurement in Grades 9–12* is granted for educational purposes. Permission must be obtained when content from this publication is used commercially, when the material is quoted in advertising, when portions are used in other publications, or when charges for copies are made. The use of material from *Navigating through Measurement in Grades 9–12*, other than in those cases described, should be brought to the attention of the National Council of Teachers of Mathematics.

The contents of the CD-ROM may not be reproduced, distributed, or adapted without the written consent of NCTM, except as noted here: The blackline masters may be downloaded and reproduced for classroom distribution; the applets may be used for instructional purposes in one classroom at a time.

The publications of the National Council of Teachers of Mathematics present a variety of viewpoints. The views expressed or implied in this publication, unless otherwise noted, should not be interpreted as official positions of the Council.

Printed in the United States of America

Table of Contents

About This Book vii

Introduction 1

CHAPTER 1
The Process of Measurement 11
    Quest for the Golden Ruler 13
    Approximately Speaking 16
    Early Measuring Devices 19

CHAPTER 2
Using Formulas to Measure Complex Shapes 23
    Mathematical Goat 25
    Chip off the Old Block 28
    Cones from Cubes 31
    Measuring a Geometric Iteration 36

CHAPTER 3
Discovering and Creating Measurement Formulas .. 41
    Discovering the Volume of a Pyramid 43
    Measuring the Size of a Tree 47

CHAPTER 4
Classic Examples of Measurement 55
    If the Earth Is Round, How Big Is It? 57
    Moon Ratios 61
    How Far Is the Sun? 66

CHAPTER 5
Measuring with Advanced Technology 71
    Starbucks Expansion 73
    Golf Ball Boogie 76
    Bouncing Ball 79
    Most Like It Hot 82

Looking Back and Looking Ahead 87

APPENDIX
Blackline Masters and Solutions 89
    Counting on Commensurability 90
    That's Irrational 91
    Super Bowl Shipment 93
    Paula's Popcorn Box 94

Rounding Numbers in a Sum 95
The Right Rope 96
Why Ships Measure Speed in Knots 97
More Measurement Methods 99
Mathematical Goat 100
Chip off the Old Block 104
A Base on a Face 107
Going for the Max 108
Making a Model 110
Iterating on a Plane 112
Iterating in 3-D 115
Building Pyramids with Cubes 117
Probing Pyramids with Spreadsheets 119
Making a Formula 121
Using Your Formula 123
Reporting Your Results 126
Assessing a Poster 127
If the Earth Is Round, How Big Is It? 129
Using a Distance-to-Diameter Ratio 131
Using a Ratio of Time 133
Figuring Out the Phases 134
Angling for the Distance 137
Starbucks Expansion 140
Golf Ball Boogie 142
Bouncing Ball 145
Most Like It Hot 148

Solutions for the Blackline Masters 151

REFERENCES 173

CONTENTS OF THE CD-ROM

Applets

Pan Balance
    Activity: Pythagoras's Quandary
Spinning and Slicing Polyhedra
    Activity: Visualizing Slices of a Cube

Blackline Masters and Templates

Those listed above, plus the following:
Measuring Tape
Protractor
Grid Paper (5 squares per inch)
Centimeter Grid Paper
"Golf Ball Bounce"
Sample Data for "Bouncing Ball"

Readings from Publications of the National Council of Teachers of Mathematics

Designing the Dynamic Domino Race
    Elizabeth George Bremigan
    Mathematics Teacher

The Triangles of Aristarchus
  Alan W. Hirshfeld
Mathematics Teacher

Using Graphing Calculators to Model Real-World Data
  Berchie W. Holliday and Lauren R. Duff
Mathematics Teacher

An Error Analysis Model for Measurement
  Donald R. Kerr, Jr., and Frank K. Lester
Measurement in School Mathematics
1976 Yearbook of the National Council of Teachers of Mathematics

The Archaeological Dig Site: Using Geometry to Reconstruct the Past
  Patricia S. Moyer and Wei Shen Hsia
Mathematics Teacher

Forever May Only Be a Few Seconds
  James J. O'Connor
Mathematics Teacher

Sports and Distance-Rate-Time
  Patrick R. Perdew
Mathematics Teacher

Teaching the Logistic Function in High School
  Gregory P. Stephens
Mathematics Teacher

About This Book

Measurement is a central strand of school mathematics and indeed is recognized as such in *Principles and Standards for School Mathematics* (National Council of Teachers of Mathematics [NCTM] 2000). Measurement is a critical application of mathematics. From the simplest acts of counting to the development of complex formulas and models, measurement permeates the scientific and mathematical activities of students. In the early grades, students measure to compare things according to such attributes as height, weight, and temperature. Children first make these comparisons directly as, for example, when they lift two objects to see which one is lighter. They soon learn to compare objects indirectly, through the use of rulers, scales, thermometers, and other measurement tools. Their understanding of measurement deepens in the later grades as they explore such matters as unit conversions, measurement error, and more complex techniques for indirect measurement. In this book, we illustrate some of the key recommendations of *Principles and Standards* for measurement in grades 9–12.

Broadly speaking, we can say that measurement involves assigning numbers to attributes of objects in terms of units that might or might not be standard. If we define measurement in this very general sense, which would include even the counting of the objects in a set, we can say that some measurements are exact and others are only approximate, depending on the context. Measurements can also be evaluated for their precision and accuracy. Moreover, the method (including the use of measuring devices) of taking a measurement can be *reliable* or *unreliable*, *valid* or *invalid*, *usable* or *unusable*, depending on the protocols, tools, and purposes of the measurement.

Principles and Standards recommends that high school students understand that all real-world measurements of continuous attributes, such as length, density, temperature, and time, are approximate and that computations involving such measurements must take into account the precision and accuracy of the measurements. Students "should be able to make ... sensible judgments about the precision and accuracy of the values they report" (NCTM 2000, p. 322). A number of activities in this volume address these issues. However, a few words are warranted on the meaning of our terms, since conventional uses of these terms—even the word *measurement*—vary.

Scientific versus Mathematical Treatment of Uncertainty

Many NCTM authors have addressed the issue of uncertainty in measurement. Finley (1928), Betz (1928), Shuster (1928; 1954), Gager (1954), and Payne and Seber (1959) all contend that the more significant digits a measurement has, the more accurate it is. These authors' ideas reflect subtle differences; what Gager calls the "precision" of a measurement, Payne and Seber refer to as its "tolerance," and what the others call the "unit" of a measurement, Payne and Seber call

Many people distinguish counting from measuring and contend that measurements are always approximate. However we define measurement, its most challenging aspect at the 9–12 level is handling uncertainty in our measurements.

It would be worthwhile for the National Council of Teachers of Mathematics and the National Science Teachers Association to collaborate to standardize the language used for measurement in K–12 science and mathematics classrooms.

its "precision." Nonetheless, there is significant agreement among them that the precision of a measurement indicates its amount of possible error.

Most scientists would disagree. In scientific contexts, how precise a measurement is tends to be connected with the degree of agreement among multiple measurements of the same quantity. This book attempts to adopt terminology and meanings that are more closely aligned with this scientific usage than with traditional mathematics usage.

Conventions and terms in this volume

In both real-world and theoretical contexts, some measurements are exact and others are only approximate. The amount of money in a cash register is a measurable attribute of the cash register and can be determined exactly by counting. However, the amount of money in circulation in the United States, although a measurable attribute of the economy, can be determined only approximately, because of the sheer complexity of the counting task.

In the case of a real-world object—a string, for example—an attribute such as length can be measured only approximately, not because of the complexity of the task, but because of the continuous nature of the attribute. In theoretical or hypothetical contexts, however, we can stipulate that a segment is exactly one meter in length and thereby treat this measurement as exact in our reasoning. Thus, the number of meters in the circumference of a circle whose radius is exactly one meter is exactly 2π meters. Some measurements in theoretical contexts are only approximate, such as the area under the standard normal curve for $-1 \leq x \leq 1$, found in tables at the back of statistics textbooks.

We often derive measurements by means of mathematical models and formulas that involve simpler measurements. For example, by using shadows and similar triangles, we can obtain an indirect measurement of the height of a tree, as compared with the direct measurement that we could obtain by using a measuring tape.

Some units, by their very definition or primary means of determination, depend on other, more fundamental units. For example, units for density are derived from units for mass and volume; and units for rates of change, such as velocity or acceleration, are derived from units for length and time. In general, units that are defined in terms of the fundamental units for length, mass, temperature, time, and energy are called *derived units*, and we will sometimes refer to measurements involving derived units as *derived measurements*.

There are conventional ways to report an approximate measurement. At a hardware store, you can buy a 4-and-1/2-inch lag screw. But is the length of the lag screw in inches exactly the real number 4.5? Not if we believe the laws of probability. Length is an attribute whose measure, though bounded, can take on values from an infinite continuum. Moreover, the real number 4.5 has an infinite decimal expansion ($4.5\bar{0}$), and it is impossible to check every digit to determine whether a given lag screw has that exact infinite string of zeros in its measurement.

Of course, manufacturers of lag screws do classify some of their products as "4 1/2-inch" screws. If they follow a commonly used

It is worth noting that formal mathematics conventionally refers to the *measure*, rather than the *measurement*, of an attribute, such as the length of a segment, to avoid suggesting any physical act of measuring.

"round-off" method of reporting measurements, "4 1/2 inch" means that the length of the screw is rounded off to the nearest half inch, since that is the smallest unit reported in the measurement. The true length is assumed to be in the interval between 4 1/4 inches and 4 3/4 inches.

The important point to note is that, when we report measurements of continuous attributes such as length, conventional practice dictates that we report an interval of values within which we claim the true value lies. Because the size of this interval is an indicator of the amount of uncertainty in the measurement, scientific sources often recommend that the interval be stated explicitly and not simply implied, as in the round-off method.

The smaller the interval of possible values associated with a measurement, the more precise the measurement. Determining the size of this interval, and hence the *precision* of the value reported, can challenge students, especially when they are working with derived measurements or replicated measurements that yield different answers for the same quantity.

Scientists think of this uncertainty interval in terms of the mean and standard deviation of repeated measurements of a quantity, and they speak of precision as the degree of agreement between replicated measurements. They invoke statistical procedures, including "confidence intervals," to assess the degree of agreement.

For simplicity, many high school mathematics textbooks define precision as the smallest unit used in expressing the measurement of a quantity according to the round-off method just described. Thus, if the length of a lag screw is reported to be 4.50 inches, then the interval of possible values is presumed to be between 4.495 inches and 4.505 inches, an interval whose length is .01 inches, or the precision of the reported value. We would say that the measurement is "precise to the nearest hundredth of an inch." In cases of derived or replicated measurements, the challenge is to determine the smallest unit to use in expressing the results so that the implied uncertainty interval represents an appropriate estimate of the uncertainty in the measured value.

The term *accuracy* refers to the degree of agreement between the measured value and the true value, if it is known, or the "accepted" true value. It is important to note that the precision of a measurement tells us very little about the accuracy of the measurement. Systematic errors that are due to the measuring technique or instrument may result in very precise measurements that are nevertheless relatively far from the true value.

Many of us have had the experience of weighing ourselves on a bathroom scale that, with great precision, indicated that we had gained, say, 5 1/2 pounds, when in fact the scale had not been properly set to zero before we stepped on it. In addition, even if we trust the reported measurement, and the true (or accepted) value is within the implied interval of possible values, the accuracy of the measurement may still be limited. To understand this, we must consider how *accuracy* is defined.

Some high school mathematics textbooks define *accuracy* as the *relative error* in the measurement. We will adopt this practice. The relative error of a measurement is defined as

$$\frac{|\text{True value} - \text{Measured value}|}{\text{True value}}, \text{ or } \frac{\text{Measurement error}}{\text{True value}}.$$

Since the true value is typically not known, we estimate the accuracy of a measurement by the ratio

$$\frac{\textit{Maximum possible error in the measurement}}{\textit{Measurement}}.$$

For example, if you measure the length of an airport runway to the nearest foot, with a maximum error of measurement of 6 inches, and get 2958 feet, your measurement is not as precise as the measurement that you would obtain if you measured the length of a pencil to the nearest inch and got 5 inches, with a maximum error of measurement of a half inch. Nonetheless, we would hesitate to say that the pencil measurement is very accurate since the size of the potential error is quite large in comparison with the size of the pencil. In the case of the pencil, the accuracy is

$$\frac{\frac{1}{2} \text{ inch}}{5 \text{ inches}},$$

or 10 percent, whereas in the case of the runway, the relative error is

$$\frac{\frac{1}{2} \text{ foot}}{2958 \text{ feet}},$$

or .02 percent. The smaller the relative error, the greater the accuracy of the measurement.

The precision and accuracy of measurements depend on the tools and methods that we use in obtaining the measurements. Before we draw a conclusion about the appropriate precision to use in expressing a measurement or become absorbed in analyzing the measurement's accuracy, we should satisfy ourselves that our tools and methods are capable of providing a *valid* and *reliable* measurement.

A tool or method is *valid* if we can reasonably argue that the measure that it gives of the intended attribute suffices for our purposes. For example, though doctors might consider weighing people with their clothes on to be a valid method of measuring weight in a doctor's office, researchers might not consider it a valid method for an investigation of the effects of a diet.

A measurement tool or method is *reliable* if we can obtain the same results through replicated measurements of the same phenomenon or if the results agree closely enough to suit our purposes. For example, counting the number of steps that a person takes to walk around a field might be a satisfactory method for deciding how much fencing to purchase, but it would not be a reliable method for measuring the perimeter of a field for a land survey. Repeating the process of walking around the field, even if the same person does the walking, is likely to result in significantly different measurements. Indeed, as the above examples illustrate, questions of validity and reliability are sometimes debatable because of the varying purposes of our measurements.

Significant digits

Since there is uncertainty in all measurements of continuous quantities in real-world contexts, scientists consider as *significant* all the digits that are known in a measurement, as well as its first uncertain digit. In a reported measurement, we can determine the number of significant digits by counting from the first nonzero digit on the left, if there is one, to the last digit on the right, corresponding to the precision of the reported measurement. For example, using a meter stick graduated in 1-cm increments, a student might measure the length of a book to be 22.3 cm, estimating the last digit to the nearest 0.1 cm. The last digit, 3, is uncertain, since it is an estimate. Thus, the measurement has three significant digits.

Scientists normally report only the significant digits in a measurement, and they use scientific notation to avoid ambiguity. For example, a measurement of 4000 meters is reported as 4.00×10^3 meters when it is precise to the nearest ten meters, and as 4.000×10^3 meters when it is precise to the nearest meter.

If we use the precision of a reported measurement to estimate its maximum possible error, then we can use the number of significant digits in a measurement as a gauge of its accuracy. The more significant digits a measurement has, the greater its accuracy. For example, a conventionally reported measurement of 2034.0 meters has five significant digits, whereas a measurement of 2.034 meters has four. Although the latter measurement is more precise, the former is more accurate. The accuracy of 2034.0 is

$$\frac{.05}{2034.0},$$

or .002 percent, whereas the accuracy of 2.034 is

$$\frac{.0005}{2.034},$$

or .02 percent. The smaller the percentage, the smaller the relative error, and the greater the accuracy.

The activity Approximately Speaking in chapter 1 calls on students to find the volume of a popcorn box, given that its length, width, and height, to the nearest centimeter, are 17 cm, 8 cm, and 25 cm, respectively (see pp. 16–17, 94, and 152). This problem can serve to illustrate the use of significant digits in a calculation of derived measurements.

Students might use the rules that they learned for significant digits in a science classroom. If so, they might compute the volume as 17 cm × 8 cm × 25 cm, or 3400 cm³, and then round the product to one significant digit, since one of the factors has only one significant digit. Thus, they would report the answer as 3000 cm³, or 3×10^3 cm³, implying that the box's volume is in the interval between 2500 cm³ and 3500 cm³.

However, since the dimensions reported for the box were rounded to the nearest centimeter, the exact volume of the box could be as high as 17.5 cm × 8.5 cm × 25.5 cm, or 3793.125 cm³, or as low as 16.5 cm × 7.5 cm × 24.5 cm, or 3031.875 cm³. Note that the answer that students would obtain by using the significant-digits method (3000 cm³) is not even in this interval of possible volumes. Also observe that the interval suggested by the significant digits method (2500 cm³ to 3500 cm³) leaves out a large portion of the possible values for the volume.

However, if students simply computed the volume of the popcorn box as 17 cm × 8 cm × 25 cm and reported the volume as 3400 cm³, without any rounding, they would do even worse. Not rounding off leaves ambiguous which digits are significant and what degree of precision the measurement has. To assume that all the digits are significant, and hence that the measurement is precise to the nearest cubic centimeter, as such an answer would suggest, would imply that the box's volume lies between 3399.5 cm³ and 3400.5 cm³, a far grosser understatement of the possible values of the exact volume.

So, what are we to conclude? The rules for significant digits that students learn in science classes are useful rules of thumb for rounding answers to an appropriate precision. However, these rules are not perfect and must be seen in the light of a deeper understanding of the intervals of possible values associated with measurements. *Principles and Standards for School Mathematics* and the activities in this volume emphasize this goal. Thus, this book stresses the importance of helping students understand how to compute uncertainty intervals associated with measured values and how to report them explicitly.

In the case of the popcorn box, students might compute the midpoint between 3031.875 cm³ and 3793.125 cm³, and report the interval of possible values as 3412.5 cm³ ± 380.625 cm³. Although this method correctly identifies the interval of possible values, it is somewhat misleading insofar as it reports a measurement (3412.5 cm³) and its uncertainty (± 380.625 cm³). Since we are not even sure what the hundreds digit should be in the true volume, the digits after the hundreds digit in 3412.5 cm³ are rather meaningless. Similarly, it makes little sense to carry the estimate of the uncertainty in a measurement beyond one significant digit.

A good rule of thumb, therefore, is to advise students to round the uncertainty estimate to one significant digit and round the reported measurement to the same decimal place—in this case, the nearest hundred. Thus, students would report the volume of the box to be 3400 cm³ ± 400 cm³. If the students had to reduce this explicitly stated interval to a single value, then they could round the measurement to the nearest thousand so that the implied uncertainty would have the same order of magnitude (in this case hundreds) as that indicated by the explicitly stated interval. The resulting value, 3000 cm³, happens to be what the significant digits method suggested in the first place.

Principles and Standards for School Mathematics makes no specific recommendations regarding the rules of thumb involving significant digits. However, in the spirit of the coherence and consistency that *Principles and Standards* emphasizes in the Curriculum Principle (NCTM 2000, p. 14), it would be useful for the mathematics and science teachers in a school to coordinate their expectations in this matter.

Overview of the Chapters

This book consists of five chapters. All the chapters highlight the importance of the expectations for grades 9–12 outlined in NCTM's Measurement Standard:

- Chapter 1 presents an overview of the essential attributes of the measurement process and focuses on the fundamental issues of number and unit, uncertainty in measurement, and measurement method.
- Chapter 2 shows how simple measurement formulas in geometry can be used as a basis for solving complex measurement problems in theoretical contexts.

"The science program should be coordinated with the mathematics program to enhance student use and understanding of mathematics in the study of science and to improve student understanding of mathematics."
(National Research Council 1996, p. 214)

- Chapter 3 provides sample activities that allow students to generate measurement formulas in both theoretical and real-world contexts.
- Chapter 4 uses historical examples that show how astronomical measurements resulted from the application of simple mathematical models.
- Chapter 5 illustrates the potential of modern technology to give measurements necessary for the mathematical modeling of real-world relationships.

Chapters 2 and 3 both illustrate the power of iteration in the use and development of formulas. Chapters 4 and 5 both focus on aspects of measurement that involve mathematical modeling.

Using the book

The book features student activities that develop the main ideas through a variety of pedagogical styles. Some activities offer guided explorations that call only for paper, pencil, and calculator. Some are hands-on investigations, and some present a sequence of problem-solving challenges that lead to a final result. The text for each activity includes assessment suggestions that identify important skills, processes, and essential understandings.

Activity sheets for students appear as reproducible blackline masters in the appendix of the book, along with solutions to the problems. Many activities have more than one part, each with its own blackline activity sheet. An icon (see key) in the margin signals all blackline pages, which readers can also print from the CD-ROM that accompanies the book.

Some discussions and activities use supplementary materials found on the CD-ROM. Special computer applets complement activities and ideas in the text, and teachers can let students use the applets in conjunction with particular activities or apart from them, to extend and deepen students' understanding.

Readings for teachers' professional development also appear on the CD. A second icon in the text alerts readers to materials on the CD-ROM.

Throughout the book, margin notes supply teaching tips, suggestions about related materials on the CD-ROM, and pertinent statements from *Principles and Standards for School Mathematics*. A third icon alerts the reader to these statements, which highlight relevant expectations for students in the area of measurement, as articulated in NCTM's Measurement Standard.

Key to Icons

Blackline Master

CD-ROM

Principles and Standards

Three different icons appear in the book, as shown in the key. One signals the blackline masters and indicates their locations in the appendix, another points readers to supplementary materials on the CD-ROM that accompanies the book, and a third alerts readers to material quoted from *Principles and Standards for School Mathematics*.

NAVIGATIONS SERIES

GRADES 9–12

NAVIGATING through MEASUREMENT

Introduction

Measurement is one of the most fundamental of all mathematical processes, permeating not only all branches of mathematics but many kindred disciplines and everyday activities as well. It is an area of study that must begin early and continue to develop in depth and sophistication throughout all levels of learning.

In its most basic form, measurement is the assignment of a numerical value to an attribute or characteristic of an object. Familiar elementary examples of measurements include the lengths, weights, and temperatures of physical things. Some more advanced examples might include the volumes of sounds or the intensities of earthquakes. Whatever the context, measurement is indispensable to the study of number, geometry, statistics, and other branches of mathematics. It is an essential link between mathematics and science, art, social studies, and other disciplines, and it is pervasive in daily activities, from buying bananas or new carpet to charting the heights of growing children on the pantry doorframe or logging the gas consumption of the family automobile. Throughout the pre-K–12 mathematics curriculum, students need to develop an understanding of measurement concepts that increases in depth and breadth as the students progress. Moreover, they need to become proficient in using measurement tools and applying measurement techniques and formulas in a wide variety of situations.

Components of the Measurement Standard

Principles and Standards for School Mathematics (NCTM 2000) summarizes these requirements, calling for instructional programs from prekindergarten through grade 12 that will enable all students to—

- understand measurable attributes of objects and the units, systems, and processes of measurement; and
- apply appropriate techniques, tools, and formulas to determine measurements.

Understanding measurable attributes of objects and the units, systems, and processes of measurement

Measurable attributes are quantifiable characteristics of objects. Recognizing which attributes of physical objects are measurable is the starting point for studying measurement, and very young children begin their exploration of measurable attributes by looking at, touching, and comparing physical things directly. They might pick up two books to see which is heavier or lay two jump ropes side by side to see which is longer. Parents and teachers have numerous opportunities to help children develop and reinforce this fundamental understanding by asking them to pick out the smallest ball or the longest bat or to line up the teddy bears from shortest to tallest. As children develop an understanding of measurement concepts, they should simultaneously develop the vocabulary to describe them. In the early years, children should have experience with different measurable attributes, such as weight (exploring *heavier* and *lighter*, for example), temperature (*warmer* and *cooler*), or capacity (discerning the glass with the *most* milk, for instance), but the emphasis in the early grades should be on length and linear measurements.

As children measure length by direct comparison—placing two crayons side by side to see which is longer, for example—they learn that they must align the objects at one end. Later, they learn to measure objects by using various units, such as a row of paper clips laid end to end. They might compare each of several crayons to the row and use the results to decide which crayon is longest or shortest. Another time, they might use a row of jumbo paper clips to measure the same crayons, discovering in the process that the size of the measuring unit determines how many of those units they need. Their experiences also should lead them to discover that some units are more appropriate than others for a particular measurement task—that, for example, paper clips may be fine for measuring the lengths of crayons, but they are not practical for measuring the length of a classroom. As their experience with measuring things grows, students should be introduced to standard measuring units and tools, including rulers marked in inches or centimeters.

Children in prekindergarten through grade 2 should have similar hands-on experiences to lay a foundation for other measurement concepts. Such experiences should include using balance scales to compare the weights of objects, filling various containers with sand or water and transferring their contents to containers of different sizes and shapes to explore volume; and working with fundamental concepts of time and learning how time is measured in minutes, hours, days, and so forth—although actually learning to tell time may wait until the children are a bit older. By the end of the pre-K–2 grade band, children should understand that the fundamental process of measurement is to identify a measurable attribute of an object, select a unit, compare that unit to the object, and report the number of units. In addition, they should have had ample opportunities to apply that process through hands-on activities involving both standard and nonstandard units, especially in measuring lengths.

As children move into grades 3–5, their understanding of measurement deepens and expands to include the measurement of other attributes, such as angle size and surface area. They learn that different kinds of units are needed to measure different attributes. They realize, for example, that measuring area requires a unit that can cover a surface, whereas measuring volume requires a unit that can fill a three-dimensional space. Again, they frequently begin to develop their understanding by using convenient nonstandard units, such as index cards for covering the surface of their desks and measuring the area. These investigations teach them that an important attribute of any unit of area is the capacity to cover the surface without gaps or overlaps. Thus, they learn that rectangular index cards can work well for measuring area, but circular objects, such as CDs, are not good choices. Eventually, the children also come to appreciate the value of standard units, and they learn to recognize and use such units as a square inch and square centimeter.

Instruction during grades 3–5 places more emphasis on developing familiarity with standard units in both customary (English) and metric systems, and students should develop mental images or benchmarks that allow them to compare measurements in the two systems. Although students at this level do not need to make precise conversions between customary and metric measurements, they should form ideas about relationships between units in the two systems, such as that one centimeter is a little shorter than half an inch, that one meter is a little longer than one yard or three feet, that one liter is a little more than one quart, and that one kilogram is a little more than two pounds. They should also develop an understanding of relationships within each system of measurement (such as that twelve inches equal one foot or that one gallon is equivalent to four quarts). In addition, they should learn that units within the metric system are related by factors of ten (e.g., one centimeter equals ten millimeters, and one meter equals one hundred centimeters or one thousand millimeters). Students should clearly understand that in reporting measurements it is essential to give the unit as well as the numerical value—to report, for example, "The length of my pencil is 19 centimeters" (or 19 cm)—not simply 19.

During these upper elementary grades, students should also encounter the notion of precision in measurement and come to recognize that all measurements are approximations. They should have opportunities to compare measurements of the same object made by different students, discussing possible reasons for the variations. They should also consider how the chosen unit affects the precision of measurements. For example, they might measure the length of a sheet of paper with both a ruler calibrated in millimeters and a ruler calibrated only in centimeters and compare the results, discovering that the first ruler allows for a more precise approximation than the second. Moreover, they should gain experience in estimating measurements when direct comparisons are not possible—estimating, for instance, the area of an irregular shape, such as their handprint or footprint, by covering it with a transparent grid of squares, counting whole squares where possible and mentally combining partial squares to arrive at an estimate of the total area. In their discussions, they should consider how precise a measurement or estimate needs to be in different contexts.

Measurement experiences in grades 3–5 also should lead students to identify certain relationships that they can generalize to basic formulas. By using square grids to measure areas of rectangles, students might begin to see that they do not need to count every square but can instead determine the length and width of the rectangle and multiply those values. Measurement experiences should also help students recognize that the same object can have multiple measurable attributes. For example, they might measure the volume, surface area, side length, and weight of a wooden cube, expressing each measurement in the appropriate units. From the recognition that multiple attributes belong to the same object come questions about how those attributes might be related. If the side length of a cube were changed, for instance, what would be the effect on the cube's volume or its surface area? Similar questions arise in comparisons between various objects. Would two rectangles with equal perimeters necessarily have the same area? What about the converse? Would two rectangles with equal areas necessarily have the same perimeter? All these measurement lessons should help students appreciate how indispensable measurement is and how closely it is tied to number and operations, geometry, and the events of daily life.

Understanding of and proficiency with measurement should flourish in the middle grades, especially in conjunction with other parts of the mathematics curriculum. As students develop familiarity with decimal numeration and scientific notation and facility in computation with decimals, applications involving metric measurements provide a natural context for learning. As students develop proportional reasoning and learn to evaluate ratios, comparisons between measurements, such as the perimeters or areas of similar plane figures, become more meaningful. Their study of geometry requires students to measure angles as well as lengths, areas, and volumes and lets students see how measurements underlie classifications of geometric figures. For example, they identify triangles as acute, right, or obtuse by evaluating measurements of their angles or classify them as equilateral, isosceles, or scalene by comparing measurements of their sides. Proportional reasoning, geometry, and measurement converge when students create or analyze scale drawings or maps. Algebraic concepts of function that develop in the middle grades have applications in relationships such as that linking distance, velocity, and time. In science classes, students use both measurement and ratios to develop concepts such as density (the ratio of mass to volume) and to identify substances by determining their densities. Through experimentation, they discover that water freezes at 0° Celsius or 32° Fahrenheit and boils at 100° Celsius or 212° Fahrenheit, and from these data they can develop benchmarks for comparing the two scales. (For example, they can see that a ten-degree change in the Celsius temperature corresponds to an eighteen-degree change in the Fahrenheit temperature or that a forecast high temperature of 30° Celsius signals a hot day ahead.)

Middle-grades students should become proficient in converting from one unit to another within a system of measurement; they should know equivalences and convert easily among inches, feet, and yards or among seconds, minutes, hours, and days, for example. They should develop benchmarks for both customary and metric measurements that can serve as aids in estimating measurements of objects. For example, they might estimate the height of a professional basketball player as about two meters by using the approximate height of a standard doorframe as a benchmark for two meters, or they might use a right angle as a basis for approximating other angle measurements like 30, 45, or 60 degrees. Although students do more computations of measurements such as areas and volumes during the middle grades than in the earlier years, they still need frequent hands-on measurement experiences, such as tiling a surface with square tiles, making shapes on a geoboard, or building a prism with blocks or interlocking cubes, to solidify their understanding of measurement concepts and processes.

By the time students reach high school, they should be adept at using the measurement concepts, units, and instruments introduced in earlier years, and they should be well grounded in using rates, such as miles per hour or grams per cubic centimeter, to express measurements of related attributes. As they engage in measurement activities during grades 9–12, students are increasingly likely to encounter situations in which they can effectively employ powerful new technologies, such as calculator-based labs (CBLs), graphing calculators, and computers, to gather and display measurements. Such instruments can report measurements, often with impressive precision, but students do not always understand clearly what is measured or how the technology has made the measurement. How a measurement of distance is obtained when a tape measure is stretched between two points is obvious; it is not so obvious when an electronic instrument reflects a laser beam from a surface. Thus, students need a firm foundation both in measurement concepts and in how to interpret representations of measurements and data displayed on screens.

Also during the high school years, students encounter new, nonlinear scales for measurement, such as the logarithmic Richter scale used to report the intensity of earthquakes (a reading of 3 on the Richter scale signifies an earthquake with ten times the intensity of an earthquake with a Richter-scale measurement of 2). Especially in their science classes, students learn about derived units, such as the light-year (the distance that light travels in one year, moving at the rate of $3(10^8)$ meters per second, or about 186,000 miles per second) or the newton (N) (the unit of force required to give an acceleration of 1 m/sec^2 to a mass of 1 kilogram). Students also extend ideas of measurement to applications in statistics when they measure certain characteristics of a sample and use those data to estimate corresponding parameters of a population. Students preparing for a more advanced study of mathematics begin to consider smaller and smaller iterations—infinitesimals, limits, instantaneous rates of change, and other measurement concepts leading to the study of calculus.

Applying appropriate techniques, tools, and formulas to determine measurements

To learn measurement concepts, students must have hands-on experiences with concrete materials and exposure to various techniques,

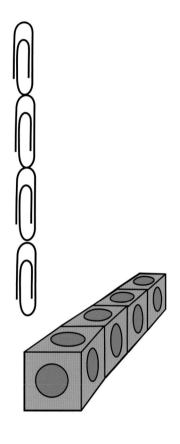

such as counting, estimating, applying formulas, and using measurement tools, including rulers, protractors, scales, clocks or stopwatches, graduated cylinders, thermometers, and electronic measuring instruments.

In the pre-K–2 years, students begin to explore measurement with a variety of nonstandard as well as standard units to help them understand the importance of having a unit for comparison. Such investigations lead them to discoveries about how different units can yield different measurements for the same attribute and why it is important to select standard units. For young children, measurement concepts, skills, and the vocabulary to describe them develop simultaneously. For example, children might learn to measure length by comparing objects to "trains" made from small cubes, discovering as they work that the cubes must be placed side by side in a straight row with no gaps, that all the cubes must be the same size (though not necessarily the same color), and that one end of the object that they want to measure must be aligned with one end of the cube train. Later, when they learn to use rulers to measure length, they must learn how to locate the zero on the ruler's scale and align it with one end of the object that they are measuring. When they attempt tasks of greater difficulty, such as measuring an attribute with a unit or instrument that is smaller than the object being measured—the width of their desks with a 12-inch ruler or a large index card, for instance—they must learn how to iterate the unit by moving the ruler or card and positioning it properly, with no gaps or overlaps from the previous position. Furthermore, they must learn to focus on the number of units and not just the numerals printed on the ruler—counting units, for example, to determine that the card shown in the illustration is three inches wide, not six inches.

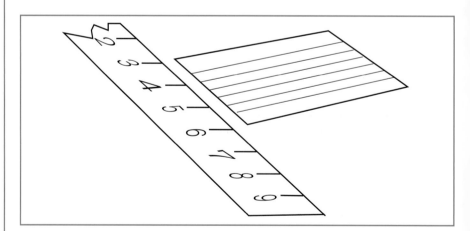

While students in prekindergarten through grade 2 are becoming acquainted with simple measuring tools and making comparisons and estimating measurements, students in grades 3–5 should be expanding their repertoires of measurement techniques and their skills in using measuring tools. In addition to becoming adept at using standard tools like rulers, protractors, scales, and clocks, third- through fifth-grade students should also encounter situations that require them to develop new techniques to accomplish measurement tasks that cannot be carried out directly with standard instruments. For example, to measure the circumference of a basketball, they might decide to wrap a string around the ball and then measure the length of the string; to measure the volume of a rock, they might submerge it in a graduated cylinder containing a known volume of water to obtain the total volume of water plus rock; to measure the weight of milk in a glass, they might weigh the empty glass as well as the glass and milk together.

As students in grades 3–5 hone their estimation skills, they should also be refining their sense of the sizes of standard units and the reasonableness of particular estimates. They might recognize 125 centimeters as a reasonable estimate for the height of a third grader but know that 125 meters or 1.25 centimeters could not be, or that a paper clip could weigh about a gram but not a kilogram. Students also should discuss estimation strategies with one another and compare the effectiveness of different approaches. In so doing, they should consider what degree of precision is required in a given situation and whether it would be better to overestimate or underestimate.

In grades 3–5, students also learn that certain measurements have special names, like *perimeter, circumference,* or *right angle*; and, as discussed earlier, they should look for patterns in measurements that will lead them to develop simple formulas, such as the formulas for the perimeter of a square, the area of a rectangle, or the volume of a cube. Through hands-on experience with objects, they should explore how different measurements might vary. For instance, by rearranging the seven tangram pieces to form a square, trapezoid, parallelogram, triangle, or nonsquare rectangle, they should find that the areas of all the shapes are the same, since they are made from the same seven pieces, but that the perimeters are different.

During middle school, students should apply their measurement skills in situations that are more complex, including problems that they can solve by decomposing or rearranging shapes. For example, they might find the area of an irregular shape on a geoboard by partitioning it into rectangles and right triangles (A) or by inscribing it in a rectangle and subtracting the areas of the surrounding shapes (B). Extending the strategy of decomposing, composing, or rearranging, students can arrive at other formulas, such as for the area of a parallelogram (C) by transforming it into a rectangle (D), or the formula for the area of a trapezoid either by decomposing it into a rectangle and two triangles (E) or by duplicating it to form a parallelogram with twice the area of the trapezoid (F). Other hands-on explorations that guide students in deriving formulas for the perimeter, area, and volume of various two- and three-dimensional shapes will ensure that these formulas are not just memorized symbols but are meaningful to them.

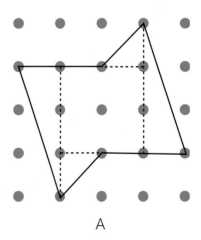

A

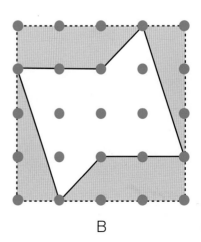

B

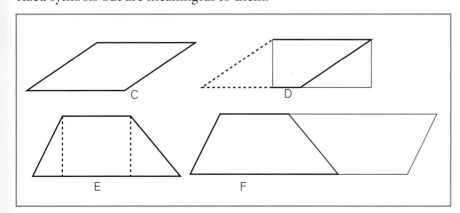

Students in grades 6-8 should become attentive to precision and error in measurement. They should understand that measurements are precise only to one-half of the smallest unit used in the measurement (for example, an angle measured with a protractor marked in degrees has a precision of ±0.5 degree, so a reported angle measurement of 52° indicates an angle between 51.5° and 52.5°). Students in the middle grades also spend a great deal of time studying ratio, proportion, and similarity—concepts that are closely tied to measurement. Students should conduct investigations of similar triangles to help them realize, for example, that corresponding angles have equal measures; that corresponding sides, altitudes, perimeters, and other linear attributes have a fixed ratio; and that the areas of the triangles have a ratio that is the square of the ratio of their corresponding sides. Likewise, in exploring similar three-dimensional shapes, students should measure and observe that corresponding sides have a constant ratio; that the surface areas are proportional to the square of the ratio of the sides; and that the volumes are proportional to the cube of the ratio of the sides.

Through investigation, students should discover how to manipulate certain measurements. For example, by holding the perimeter constant and constructing different rectangles, they should learn that the area of the rectangle will be greatest when the rectangle is a square. Conversely, by holding the area constant and constructing different rectangles, they should discover that the perimeter is smallest when the rectangle approaches a square. They can apply discoveries like these in constructing maps and scale drawings or models or in investigating how the shape of packaging, such as cracker or cereal boxes, affects the surface area and volume of the container. They also should compare measurements of attributes expressed as rates, such as unit pricing (e.g., dollars per pound or cents per minute), velocity (e.g., miles per hour [MPH] or revolutions per minute [rpm]), or density (e.g., grams per cubic centimeter). All these measurements require proportional reasoning, and they arise frequently in the middle school mathematics curriculum, in connection with such topics as the slopes of linear functions.

High school students should develop an even more sophisticated understanding of precision in measurement as well as critical judgment about the way in which measurements are reported, especially in the significant digits resulting from calculations. For example, if the side lengths of a cube were measured to the nearest millimeter and reported as 141 mm or 14.1 cm, then the actual side length lies between 14.05 cm and 14.15 cm, and the volume of the cube would correctly be said to be between 2773 cm^3 and 2834 cm^3, or (14.05 cm)^3 and (14.15 cm)^3. It would not be correct to report the volume as 2803.221 cm^3—the numerical result of calculating (14.1 cm)^3. Students in grades 9–12 also should develop a facility with units that will allow them to make necessary conversions among units, such as from feet to miles and hours to seconds in calculating a distance in miles (with the distance formula $d = v \cdot t$), when the velocity is reported in feet per second and the time is given in hours. Building on their earlier understanding that all measurements are approximations, high school students should also explore how some measurements can be estimated by a series of successively more accurate approximations. For example, finding the perimeter of inscribed and circumscribed n-gons as n increases (n = 3, 4, 5, ...) leads to approximations for the circumference of a circle.

High school students can use their mathematical knowledge and skills in developing progressively more rigorous derivations of important measurement formulas and in using those formulas in solving problems, not only in their mathematics classes but in other subjects as well. Students in grades 9–12 should apply measurement strategies and formulas to a wider range of geometric shapes, including cylinders, cones, prisms, pyramids, and spheres, and to very large measurements, such as distances in astronomy, and extremely small measurements, such as the size of an atomic nucleus or the mass of an electron. Students should also encounter highly sophisticated measurement concepts dealing with a variety of physical, technological, and cultural phenomena, including the half-life of a radioactive element, the charge on an electron, the strength of a magnetic field, and the birthrate of a population.

Measurement across the Mathematics Curriculum

A curriculum that fosters the development of the measurement concepts and skills envisioned in *Principles and Standards* needs to be coherent, developmental, focused, and well articulated. Because measurement is pervasive in the entire mathematics curriculum, as well as in other subjects, it is often taught in conjunction with other topics rather than as a topic on its own. Teaching measurement involves offering students frequent hands-on experiences with concrete objects and measuring instruments, and teachers need to ensure that students develop strong conceptual foundations before moving too quickly to formulas and unit conversions.

The *Navigating through Measurement* books reflect a vision of how selected "big ideas" of measurement and important measurement skills develop over the pre-K–12 years, but they do not attempt to articulate a complete measurement curriculum. Teachers and students who use other books in the Navigations Series will encounter many of the concepts presented in the measurement books there as well, in other contexts, in connection with the Algebra, Number, Geometry, and Data Analysis and Probability Standards. Conversely, in the *Navigating through Measurement* books, as in the classroom, concepts related to this Standard are applied and reinforced across the other strands. The four *Navigating through Measurement* books are offered as guides to help educators set a course for successful implementation of the very important Measurement Standard.

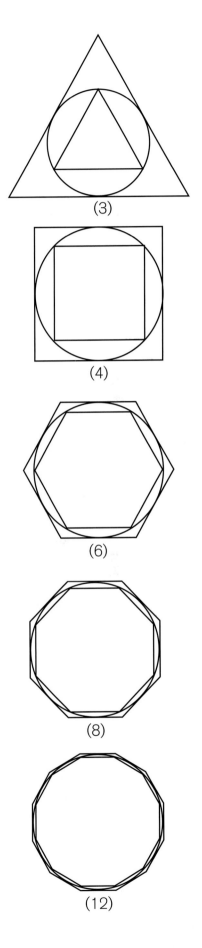

NAVIGATIONS SERIES

GRADES 9–12

NAVIGATING *through* MEASUREMENT

Chapter 1
The Process of Measurement

"Instructional programs ... should enable all students to ... understand measurable attributes of objects and the units, systems, and processes of measurement."
(NCTM 2000, p. 320)

We can only conjecture about the earliest origins of mathematics in human culture, since the first use of numbers was prehistoric. It seems likely, however, that numbers initially appeared in connection with measurements of various kinds. Numbers were almost certainly first used for counting: 5 apples, 3 pebbles, 7 fish, and so on. "Apples," "pebbles," and "fish" were units, and the numbers told how many of them someone had or saw.

Eventually, the concept of counting units was extended to measurable quantities like length, weight, and time: 5 inches, 3 ounces, 7 minutes, and so on. The fact that two 5-inch bricks had the same length as one 10-inch brick was useful information to our ancestors, but it also gave them insights into the mathematical operations of addition, subtraction, multiplication, and division. The need for more sophisticated measurements eventually led to the discovery of more sophisticated mathematics, a process that continues into the present day.

This chapter presents three activities that allow students to explore the process of measurement. This process requires the use of numbers and units. The first activity, Quest for the Golden Ruler, explores the close connection between our most important system of numbers for measuring things—the real numbers—and the idea of units of measurement. The process of measurement must also take account of the fact that real-world measurements of continuous quantities are not exact but approximations. The chapter's second activity, Approximately Speaking, examines the approximate nature of continuous measurements in the real world. The measurement process also makes use of diverse tools and methods. The third activity, Early Measurement Devices, lets students examine a variety of methods by which men and women have obtained measurements over the years.

Quest for the Golden Ruler

Goals

- Deepen understanding of the real numbers and the significance of irrational numbers in the measurement process
- Develop the idea of unit

Materials and Equipment

For each student—
- A copy of the activity sheet "Counting on Commensurability"
- A copy of the activity sheet "That's Irrational"
- Notebook paper, a pencil, and a calculator

For each group of two or three students (optional)—
- Access to a computer and the applet Pan Balance on the CD-ROM

pp. 90; 91–92

Discussion

The Pythagoreans believed that measurement had more than just practical importance. In their view, the connection between numbers and reality was so fundamental that a true understanding of the physical universe could be attained only by mastering the connection. Unfortunately, the Pythagoreans' assumptions about number and units of measurement gave them no theoretical way of dealing with measurements that involved irrational numbers. It is ironic that the famous right triangle theorem bearing Pythagoras's name presented that very dilemma. The activity Quest for the Golden Ruler explores the dilemma in two parts.

Part 1—"Counting on Commensurability"

The Pythagoreans believed that in theory someone who was given any two line segments and a blank ruler could divide the ruler into units capable of measuring both segments exactly with natural numbers. They treated this intuitively plausible *commensurability property* as a physical, concrete consequence of how numbers behave in the abstract—indeed, as rational numbers do behave. The first part of the activity, "Counting on Commensurability," introduces the idea of commensurability and examines why it appears to be a logical assumption.

Two quantities are *commensurable* if each quantity is exactly a whole number of units when the two quantities are measured with a common unit.

Part 2—"That's Irrational"

In the second part of the activity, students make use of the Pythagorean theorem to discover segments that are not commensurable. The proof of the premise that a leg of an isosceles right triangle is not commensurable with the hypotenuse was known to Aristotle and Euclid. The argument presented in the activity proves that there is no rational number $\frac{m}{n}$ in lowest terms such that $\left(\frac{m}{n}\right)^2 = 2$.

In modern terms, this argument leads to the discovery that $\sqrt{2}$ is an irrational number. This conclusion, however, was not available to

Your students can work with the applet Pan Balance on the accompanying CD as a preliminary investigation of commensurability.

Chapter 1: The Process of Measurement

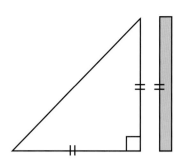

According to the NCTM Teaching Principle, "If students are to learn to make conjectures, experiment with various approaches to solving problems, construct mathematical arguments, and respond to others' arguments, then creating an environment that fosters these kinds of activities is essential."

(NCTM 2000, p. 18)

mathematicians at the time of Euclid and Aristotle, whose contemporaries had no concept of irrational numbers.

Students can extend the results of this activity by proving that the square roots of all natural numbers are irrational except for natural numbers that are perfect squares. First, ask them to convince themselves of the following facts about rational numbers:

1. If $\frac{m}{n} > 0$ is a rational number in lowest terms, then $\frac{m^2}{n^2}$ is also in lowest terms.

2. If $\frac{m}{n}$ is a rational number in lowest terms and $\frac{m}{n} = k$, a natural number, then $n = 1$.

Next, tell them that you are thinking of a natural number N such that $\sqrt{N} = \frac{p}{q}$, a rational number in lowest terms. Ask them to use facts 1 and 2 to draw some conclusions that will narrow down the possible values of N. These facts together with the assumptions about N lead immediately to the following conclusions:

(a) $N = \frac{p^2}{q^2}$;

(b) $\frac{p^2}{q^2}$ is in lowest terms;

(c) $q^2 = 1$; and

(d) $N = p^2$.

Thus, the only possible values of N that have rational square roots are the perfect squares: 1, 4, 9, 16, and so forth.

Assessment

This activity divides the important skills and processes into a sequence of steps. You will need to determine whether or not each step makes sense to your students. Furthermore, to assess your students' overall understanding of the ideas in this activity, you could have them participate in a classroom discussion. As an alternative, you could ask them to write a journal entry, explaining what it means to say that the lengths 3/5 meter and 2/7 meter are commensurable, but the lengths 3/5 meter and √5 meters are not.

Where to Go Next in Instruction

Although the irrational numbers continue to hold mysteries for present-day mathematicians, the development of irrational numbers resolved the dilemma of incommensurability. Our modern concept of real numbers exactly ensures the correspondence between number and measurement that the Pythagoreans had expected with the natural numbers.

However, even though the real number line is adequate for measuring continuous quantities, all such measurements in the real world are approximate. When we measure quantities in real-life applications and use those measurements to derive other measurements, we must account for this inevitable "fuzziness." The following activity introduces this topic at a basic level.

Approximately Speaking

Goals

Develop understanding of the approximate nature of measurements
Learn to estimate the potential error in a measurement

Materials and Equipment

For each student—
- A copy of the activity sheet "Super Bowl Shipment"
- A copy of the activity sheet "Paula's Popcorn Box"
- A copy of the activity sheet "Rounding Numbers in a Sum"
- Notebook paper, a pencil, and a calculator

For each group of two or three students (optional for assessment)—
- A ruler calibrated in centimeters
- A cardboard box (empty cereal or cracker box, for example), *or*
- Construction paper, scissors, and tape

Discussion

The activity Approximately Speaking has three parts: "Super Bowl Shipment," "Paula's Popcorn Box," and "Rounding Numbers in a Sum." Each part invites the students to explore implications of the fact that measurements of continuous quantities are approximations.

Part 1—"Super Bowl Shipment"

In part 1 of Approximately Speaking, students assess the impact of round-off error in a real-world situation involving shipment weights. A trucker has received a last-minute request to add a shipment of 100 cases of souvenir dolls for the Super Bowl to the load on his truck. He knows that he cannot add more than 8,000 pounds to his load without running a risk of failing a weight inspection. He has been told that each doll weighs 8.4 ounces and that each case contains 150 dolls plus a pound of packing material. Students investigate whether or not the trucker should accept the extra load.

This investigation should help students understand that all real-world measurements of continuous attributes, such as length, are estimates. Equally important, they should see that the uncertainty inherent in the estimates must be considered when interpreting the results of computations that use those estimates. This obvious fact is easy to forget, especially in our age of electronic calculations. Even mathematics textbooks sometimes get carried away, as, for example, when they include problems that state that someone has leaned an 8-foot ladder against a wall 5 feet away and then invite students to use the Pythagorean theorem to deduce that the top of the ladder is 6.245 feet above the ground.

Part 2—"Paula's Popcorn Box"

Part 2 of Approximately Speaking calls on students to compute the volume of a box. A student, Paula, is designing a popcorn box as a homework project. She rounds each of the dimensions of her box to the

pp. 93, 94, 95

"All students should ... analyze precision, accuracy, and approximate error in measurement situations."

(NCTM 2000. p. 320)

Rounding the results of one or more intermediate steps in a computation can produce *round-off error,* which leads to a final result that is different from the result that exact numbers would have given.

nearest centimeter and reports its dimensions as $25 \times 17 \times 8$ cm^3. This problem gives a good opening for discussing ideas about significant digits that your students have learned in their science classes.

Paula incorrectly supposes that measuring each dimension of her box to the nearest centimeter guarantees a calculation of the box's volume that is precise to the nearest cubic centimeter. The activity lets students discover that even in the case of Paula's modest-sized popcorn box, the potential error in Paula's method amounts to a pretty good handful of popcorn.

It is of course important that students understand how to compute the intervals of possible values associated with measurements, but they should also examine their intuitions about the likelihood of the values in those intervals. For example, in the case of the Super Bowl shipment, don't be surprised if they argue that the driver's luck would have to be very bad for every doll and every case to turn out to be heavier than expected, and they would have a very good point. It is far more likely that imperfections in the manufacturing process would cause some dolls to be heavier and some to be lighter, with the net effect of the differences canceling each other out. Statisticians count on this effect when they make their calculations, and they can even predict the likelihood of certain values with the *central limit theorem.*

Part 3—"Rounding Numbers in a Sum"

Part 3 of Approximately Speaking offers some everyday applications of the central limit theorem. You should discuss and perhaps even simulate these examples with your students to help them understand how the ideas work.

Assessment

This activity divides the relevant skills and processes into sequences of steps in three different parts. You will need to decide whether or not each step makes sense to your students. Check that they can compute the upper and lower bounds of the possible values when they are using rounded-off measurements to derive a measurement such as volume.

To assess students' understanding of the essential ideas presented in this activity, you might assign your students to groups, give each group a cardboard box to measure, and challenge the group members to design a method for reporting the box's measurements. Ask the students to compute the intervals of the possible values that are associated with their measurements. Have them report the resulting margin of error in their derived measurement of the volume of their box.

As an alternative, you could distribute construction paper, tape, and scissors to your students and ask them to design, construct, and measure their own boxes. This activity will engage them in exploring matters related to accuracy as well as precision as they work to square their boxes and keep their dimensions uniform at the top and bottom, and so on.

The students' designs should take account of the appropriate precision and interval of possible values of the reported measurement. As a culminating activity, you could have your class discuss the relative merits of the groups' designs.

See "About This Book" (pp. viii–xii) for a discussion of the connections among precision, accuracy, *and* significant digits *in measurements. This discussion considers possible ways of reporting the volume of Paula's popcorn box.*

Statisticians use the *central limit theorem* to explain why it is that many distributions are remarkably close to the normal distribution. If we draw many samples of size *n* from a population with mean µ and variance σ², then the distribution of the means of our sample populations will be approximately normal, with mean µ and variance σ²/*n*. The larger our number *n* is, the closer our distribution of means will be to the normal distribution.

See Navigating through Data Analysis in Grades 9–12 *(Burrill et al. 2003) for a discussion of statistical methods. Variance and how to compute it are explained in "Measures of Variability— Mean Absolute Deviation and Standard Deviation" on the CD-ROM that*

According to the NCTM Assessment Principle, "Assessment should reflect the mathematics that all students need to know and be able to do, and it should focus on students' understanding as well as their procedural skills." (NCTM 2000, p. 23)

Where to Go Next in Instruction

The precision and accuracy of our measurements depend on the measuring tools and the methods that we use. In studying the process of measurement, students need to examine the attributes of measuring tools and measuring methods. This is especially important in situations where the tools and methods are not standard. The following activity, Early Measuring Devices, introduces students to several such situations.

Early Measuring Devices

Goals

- Explore examples from the history of measurement
- Convert rates from nonstandard units to standard units
- Develop methods for measuring certain quantities reliably with simple materials

Materials and Equipment

For each student—
- A copy of the activity sheet "The Right Rope"
- A copy of the activity sheet "Why Ships Measure Speed in Knots"
- A copy of the activity sheet "More Measurement Methods"
- Notebook paper, a pencil, and a calculator

For each group of two or three students (optional for part 3—"More Measurement Methods")—
- Materials to simulate the measurement devices used in the historical examples:
  - A ruler
  - A sheet of standard-sized paper (8 1/2 by 11 inches)
  - A 3-by-5-inch index card
  - About 10 pennies and 10 nickels
  - A long rope or string (approximately 25 feet)

pp. 96; 97–98; 99

Discussion

In an age in which thermometers, speedometers, barometers, and chronometers (better known as clocks) are commonplace, we frequently take accurate measurements for granted. We consider a watch defective if it gains a few minutes per week. Students benefit enormously from the easy access to accurate measurements that modern devices provide, yet they can also benefit from learning about the many mathematically clever methods that their ancestors developed to obtain workable measurements without sophisticated technology. The explorations in this activity are designed to motivate students to think about a few of these ancient measurement techniques. (Chapter 4 pursues this topic, focusing on early astronomical measurements.)

The activity unfolds in three parts. "The Right Rope" allows students to explore the measurement possibilities offered by a rope knotted into twelve equal segments. "Why Ships Measure Speed in Knots" shows the role that a much longer knotted rope came to play in measuring the speed of a ship at sea. The rope devices in the first two parts of Early Measurement Devices are actual primitive measuring tools, as is the "glass container of sand" in part 2. Students will readily recognize this device as an hourglass. Part 3, "More Measurement Methods," presents a variety of "low-tech" applications of mathematics used in the past to obtain workable measurements indirectly.

Dauben (1992) offers additional ideas and information on measuring with ropes. See especially "Learning the Ropes: The Origins of Geometry" (pp. 2–3) and "Indian Rope Tricks: A Number of Problems" (pp. 4–5).

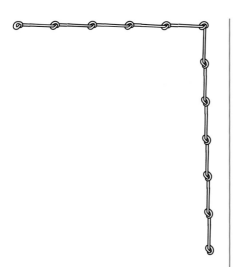

"More Measurement Methods" asks students how they could measure their distance from an approaching thunderstorm. For a discussion of this and other questions about measurements of weather, see "Weather Trivia," provided by meteorologist Len Randolph on the Web at http://www.uen.org/ weather/html/triv.html.

"Instructional programs … should enable all students to … apply appropriate techniques, tools, and formulas to determine measurements." (NCTM 2000, p. 320)

Part 1—"The Right Rope"

In part 1, students apply their knowledge of geometry to explore the benefit of measuring with a rope knotted into twelve equal parts. They discover that such a rope could conveniently form a 3-4-5 right triangle. Students consider the value of this property and the advantages and disadvantages of using the resulting right angle as a measuring tool in construction projects.

Part 2—"Why Ships Measure Speed in Knots"

Part 2 demonstrates that the word *knot*, which designates a standard unit of speed at sea, is etymologically linked to a knot in a rope. A nautical mile (technically defined as the length of 1 minute of arc on the curved surface of the earth) is equal to 6076 feet, or approximately 1.15 statute miles. Thus, the measure provided by the rope in the activity (47.25 feet per 28 seconds) is almost exactly one nautical mile per hour, as the students learn by doing the calculations.

Part 3—"More Measurement Methods"

The last part of the activity challenges students to find some useful measurements in creative but unfamiliar ways. The questions prompt students to "reinvent" some of the mathematics-based strategies that men and women devised in the past to measure quantities with simple instruments. If you wish, you can arrange your students in groups and provide them with the items suggested in the materials list so that they can carry out actual measurements of the types called for in the questions.

Students might well come up with ideas that are different from the approaches that we suggest in the solutions. The more ideas, the merrier! Encourage your students to discuss the reliability and validity of each of their methods and measuring tools.

Assessment

You should evaluate several basic skills and processes as students complete the steps of the activity. For example, you can assess your students' abilities to compute rates of change by observing the ease or difficulty with which they find the speed of a ship in knots per second. In addition, you can note whether or not they are comfortable in making conversions from one system of units to another (such as converting knots per second into miles per hour).

It is also important that your students understand the strengths and weaknesses of nonstandard units and nonstandard methods of measurement. They should be able to compare these units and methods with standard ones. Moreover, students should appreciate when methods are practical, valid, and reliable.

You can assess your students' understanding of these ideas by having them work in groups to develop a poster that presents and evaluates a measurement method that uses nonstandard units. For example, in some science classes, students learn the "fist" method for measuring angles of elevation. Using this method, students might measure the elevation of the sun, for instance. They would stand with their arms fully extended in front of them and their hands balled into fists. Then they would carefully count the number of fists, one stacked squarely on top of the other, which they needed to trace the path of their eye from the horizon to the sun. Finally, they would convert their measurements to degrees by using a conversion rule based on the number of fists they needed to sweep out an angle of 180°.

In evaluating any such method, students should present data on the precision and accuracy of the measurements. In addition, they should discuss the reliability of the method. Students' posters should also list situations and purposes for which the method would be valid or invalid.

Conclusion

Activities in this chapter have investigated ideas that are fundamental to the process of measurement as it has developed through history. The first activity stressed the importance of the real numbers in the measurement of continuous quantities. The second activity underscored the fact that all such measurements are approximations that are precise only within particular intervals. The third and final activity considered devices and methods that, although primitive, are very much a part of the history of measurement—and hence, mathematics.

Throughout history, progress in achieving more accurate measurements has always gone hand-in-hand with scientific discovery. This relationship between progress in science and advances in reliable tools and methods persists to the present day as we strive to measure increasingly complex phenomena from the farthest reaches of space as well as within the innermost structures of matter.

Chapter 2 illustrates the use of basic formulas to achieve some seemingly complex measurements. Its activities explore the application of basic measurements in a variety of mathematical contexts.

"Sports and Distance-Rate-Time" (Perdew 2002; available on the CD-ROM) presents an activity to help students deepen their understanding of converting from one rate of change to another.

NAVIGATIONS SERIES

GRADES 9–12

NAVIGATING through MEASUREMENT

Chapter 2
Using Formulas to Measure Complex Shapes

According to the NCTM Learning Principle, "When challenged with appropriately chosen tasks, students become confident in their ability to tackle difficult problems, eager to figure things out on their own, flexible in exploring mathematical ideas and trying alternative solution paths, and willing to persevere. Effective learners recognize the importance of reflecting on their thinking and learning from their mistakes." (NCTM 2000, p. 21)

The square and the cube are perhaps the most familiar shapes in two and three dimensions, respectively. They are the fundamental units that we use for measuring area and volume. This chapter presents activities that use these shapes as points of departure for measuring attributes of other, more complex shapes.

As *Principles and Standards for School Mathematics* asserts, "Measurement concepts should grow in sophistication and breadth across the grades" (NCTM 2000, p. 44). In working through the activities in this chapter, your students will derive measurements for perimeter, area, and volume. As they work, they will develop ideas that challenge their mathematical knowledge, visualization skills, and problem-solving abilities.

The chapter includes four activities: Mathematical Goat, Chip off the Old Block, Cones and Cubes, and Measuring a Geometric Iteration. Each activity guides students in using simple shapes and formulas to obtain complex measurements. The activities are intended to foster a mathematical curiosity that will impel students to probe further. Solving these isolated problems may inspire them to explore what happens when conditions change and new, related problems confront them.

In Mathematical Goat, students create, change, and compare the areas of different grazing regions for a goat. The goat's grazing region starts as a square pen, but its shape becomes increasingly complex through the owner's addition of parts of circles. Chip off the Old Block lets students explore a woodcarver's options for cutting a sphere and a cylinder of greatest volumes from given, equal-sized wooden cubes. Cones and Cubes extends this activity to cones, with possibly surprising results, and an optional supplement allows students to make a hands-on model of the situation. In Measuring a Geometric Iteration, students investigate the changing measurements of a complex structure that emerges from a simple square through an iterative process.

Mathematical Goat

Goals

- Review and apply the area formulas for squares, circles, and triangles
- Visualize and generate complex geometric shapes based on squares and circles
- Establish and compute measurement relationships

Materials and Equipment

For each student—
- Notebook paper, a pencil, and a calculator
- A copy of the activity sheet "Mathematical Goat"
- A sheet of grid paper

For each pair of students—
- A ruler or other straight edge
- A compass
- A protractor

Discussion

No doubt you have encountered or used many area problems that call on students to break down complicated regions into simple parts for easier computation. This activity alters the pattern slightly by starting with a simple problem involving a square and systematically modifying it, step by step, through a scenario about a goat in a grazing area.

An owner grazes a goat in a 12-foot square pen. After the goat has eaten all the grass inside the pen, the owner takes the goat outside the pen. The owner tethers the goat to the pen's corner post at P, first with an 8-foot rope (see fig. 2.1a), and then with a 14-foot rope (see fig. 2.1b). The goat can range to the full extent of each rope, grazing in new regions outside the old pen. Ultimately, the owner tethers the goat to each of the corner posts in turn, thus defining a very large and complex grazing region for the goat (see fig 2.1c).

pp. 100–103

You can print grid paper suitable for the activity Mathematical Goat from the template "Grid Paper (5 Squares per Inch)."

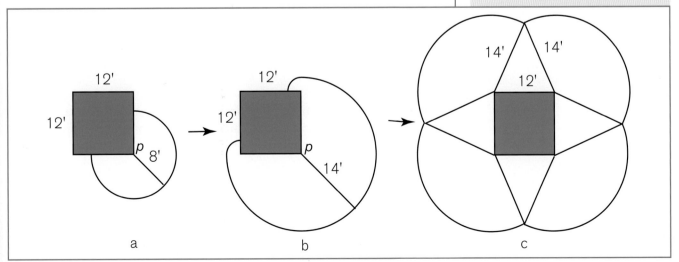

Fig. **2.1.**

Changing grazing regions for a goat

Chapter 2: Using Formulas to Measure Complex Shapes

Students can work on the activity independently or in pairs, visualizing the changing areas and constructing various parts of circles and making repeated use of area formulas. The students check the reasonableness of their area computations by counting squares that the goat's new grazing regions cover on a grid. The activity eventually leads them to construct a more complex region consisting of four congruent sectors of circles alternating with four congruent isosceles triangles.

You should encourage your students to visualize each new situation and to make careful use of a compass to draw accurate constructions at each stage. The final grazing region may defy their analytical abilities, and they may need to fall back on their earlier counting procedure as a first stab at an area solution.

The activity's final question moves students from what might be regarded as real-world applications of mathematics to a purely theoretical consideration. Here students must describe the approximate shape of a theoretical grazing region defined by a hypothetical goat on a mile-long tether that the owner moves continually from one corner post to the next.

If your students have a difficult time visualizing this final situation, you might have them use a sheet of grid paper to construct, to an appropriate scale, the grazing region for a goat on a rope whose length is 400 feet. This should help them see that with a rope one mile long, the relatively tiny 12-foot square essentially becomes the center of a circular grazing region with a radius of one mile.

Assessment

As your students work, check to see that they correctly identify the different fractional parts of the circles formed as the grazing region changes from one situation to another. The identification can be tricky, but it is crucial, since students must use geometric information in the diagrams that they construct to calculate the central angles of some sectors. If the students don't know how to find the angles in an isosceles triangle when they are given only the measurements of its sides, then you should urge them to use a protractor to measure the appropriate angles.

Also assess the students' abilities to find the area of an isosceles triangle from the lengths of its sides. (This can involve the Pythagorean theorem.) Evaluate their facility in using the formula for the area of a circle to find the area of a sector of the circle. Make sure that your students have a fundamental understanding of the ideas that the area of a complex figure is the sum of the areas of its nonoverlapping parts and that they can construct complex figures by using only circles, squares, and triangles.

You might assess your students' understanding of this last idea by challenging them to make up their own problem. Tell them that they should use only squares, circles, and triangles to produce a complicated shape whose area they will measure either by computing or by counting squares on a grid.

Where to Go Next in Instruction

Mathematical Goat relies on extensive use of area formulas and the derivation of angles and lengths. The activity gives students practice with the formulas, just as the Measurement Standard in *Principles and Standards* recommends.

The Measurement Standard also emphasizes that students in grades 9–12 should "understand and use formulas for … [the] volume of geometric figures, including cones, spheres, and cylinders" (NCTM 2000, p. 320). The following two activities, Chip off the Old Block and Cones from Cubes, illustrate how your students can satisfy this expectation at different levels of difficulty by taking a cube and considering in sequence the cylinder, sphere, and cone—all of maximum volume—that someone could cut from the cube.

Chip off the Old Block

Goals

- Review formulas for the volume and the surface area of a sphere and a cylinder
- Visualize geometric relationships among solids
- Compute and relate volumes and surface areas of cubes, spheres, and cylinders

Materials and Equipment

For each student—
- A copy of the activity sheet "Chip off the Old Block"
- Notebook paper, a pencil, and a calculator

For each group of two or three students (optional)—
- Access to a software utility for constructing and measuring geometric shapes and doing calculations

For the classroom (optional)—
- Models of a cube, a cylinder, and a sphere (with dimensions in the same relationship as in the activity: a cube of side n, a cylinder of height n and a base with radius $n/2$, and a sphere of radius $n/2$)

Discussion

Those who have stacked sets of identical square objects (floor tiles, for example) and sets of identical circular objects (CDs, for instance) can readily associate cubes with squares and cylinders with circles. Yet, we can also relate circles to squares, by inscribing or circumscribing them. A similar relationship exists in three dimensions between cylinders and cubes and between spheres and cubes. Chip off the Old Block uses this fact to help students review volume formulas. By asking students to imagine that a woodcarver needs to cut a sphere and a cylinder, both of maximum volume, from two identical wooden cubes (see fig. 2.2), the activity sets the stage for the more challenging activity that follows on cones.

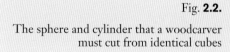

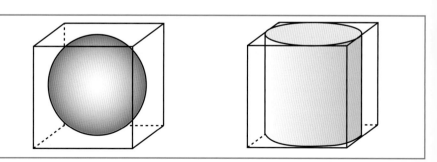

Fig. **2.2.**
The sphere and cylinder that a woodcarver must cut from identical cubes

Chip off the Old Block leads the students step by step to discoveries that are likely to surprise them. First, they will find an exact 3/2 ratio between the volumes of the cylinder and the sphere that the woodcarver cuts. Then their investigation will show them that there is also a

3/2 ratio between the surface areas of the woodcarver's cylinder and sphere. Finally, the students will discover that they can generalize their results for a sphere of radius $a/2$ and a cube of height a and radius $a/2$.

Remind your students to make their computations algebraically, expressing their answers in terms of π, as the activity suggests. They should give decimal approximations only at the final stage of a calculation, thus reinforcing their understanding of the meaning of the symbol π.

There are many other good reasons for having your students give their results in terms of π, not the least of which is to reenact a bit of mathematics history. As students compare the volumes and surface areas of the woodcarver's new solids, they work with ratios and percentages. When they express their ratios in terms of π, they can simplify their work readily and exactly to discover the somewhat surprising 3/2 ratios. Indeed, Archimedes was so pleased to discover the equality of these ratios for spheres and cylinders that he asked that the result be etched on his tombstone!

Your advanced students can prove that these results are true in general by using algebra and the formulas for volume and surface area of spheres and cylinders. The margin shows these computations.

There are many interesting ways to extend the problems in this activity. For example, you might ask your students to consider the sphere that the woodcarver cuts from a 6-centimeter cube and find the cube of largest volume that he could cut from this sphere.

If your students would welcome an even more challenging extension, ask them if the cylinder of maximum volume that can be cut from a sphere is the same as the cylinder of maximum surface area. This problem can be conveniently reduced to inscribing a rectangle in a circle, as in figure 2.3.

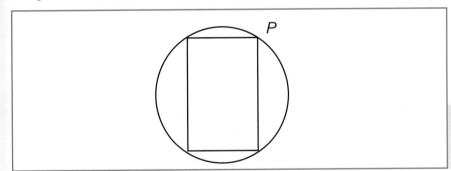

For any cylinder and sphere in the same relation to each other as in the activity (that is, the radii are both r, and the height of the cylinder is $2r$), the volume of the cylinder will be in a 3/2 ratio to that of the sphere, and the surface area of the cylinder will also be in a 3/2 ratio to that of the sphere:

The volume of any cylinder with radius r and height $2r$ is

$$\pi r^2 (2r) = 2\pi r^3.$$

The volume of any sphere with radius r is

$$(4/3)\pi r^3.$$

So the ratio of the volume of the cylinder to that of the sphere is

$$2\pi r^3/(4/3)\pi r^3 = 2/(4/3)$$
$$= 2(3/4) = 3/2.$$

The surface area of any cylinder with radius r and height $2r$ is

$$2\pi r^2 + 4\pi r^2 = 6\pi r^2.$$

The surface area of any sphere with radius r is $4\pi r^2$.

So the ratio of the surface area of the cylinder to that of the sphere is

$$6\pi r^2/4\pi r^2 = 6/4 = 3/2.$$

Fig. **2.3.**

A side view of a cylinder inscribed in a sphere

Using a software utility such as The Geometer's Sketchpad, students can compute the volumes and surface areas of the corresponding cylinders as the point P is moved on the circle. Calculus students can use an analytic approach.

Assessment

Chip off the Old Block reinforces skills and processes that are essential for students to master. Assess your students' abilities to use the formulas for the surface area and the volume of a sphere and a cylinder in their exact forms—with π—and their skills in simplifying ratios.

Also make sure that your students recognize and are capable of visualizing the limits on the dimensions of the geometric figures in the activity. For example, students should understand that the largest circle

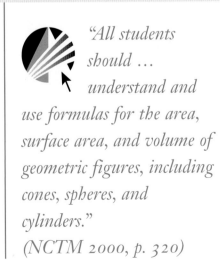

"All students should ... understand and use formulas for the area, surface area, and volume of geometric figures, including cones, spheres, and cylinders."

(NCTM 2000, p. 320)

According to the ancient historian Plutarch, Archimedes asked his friends and relatives to place over his tomb a cylinder containing a sphere, with an inscription of the ratio of the containing solid to the contained solid.

that can be inscribed in a 6-centimeter square has a radius of 3 centimeters.

Students should appreciate Archimedes' pleasure in his accomplishment and see it as fully justified. They should understand why Archimedes requested an etching of a sphere enclosed in a cylinder, along with the inscription of the ratio 3/2, on his tomb. You can assess your students' understanding of Archimedes' results by asking them to explain, in a journal entry or as a homework problem, whether or not it is unusual for two objects' volumes to be in the same ratio as their surface areas.

Where to Go Next in Instruction

In Chip off the Old Block, the most obvious positions and orientations of the cylinder and sphere within the block of wood yield the maximum volumes. This is not always the case, as the next activity shows.

Cones from Cubes

Goals

- Review the formula for the volume of a cone
- Visualize geometric relationships
- Determine dimensions from related figures

Materials and Equipment

For each student—
- A copy of the activity sheet "A Base on a Face"
- A copy of the activity sheet "Going for the Max" (optional)
- Notebook paper, a pencil, and a calculator
- Scissors, tape, a metric ruler, and a copy of the activity sheet "Making a Model" (optional)

For each group of two or three students (optional)—
- Access to the applet Spinning and Slicing Polyhedra

pp. 107; 108–9; 110–11

Discussion

Cones from Cubes extends students' work in Chip off the Old Block to the more complex situation of cutting a cone of maximum volume from a 6-centimeter cube. How should a woodcarver place the cone in the cube to maximize the cone's volume? Should the woodcarver locate the base of the cone on a face of the cube, as for the cylinder in Chip off the Old Block?

Part 1—"A Base on a Face"

Given the choice, most students would probably advise placing the cone's circular base on a face of the cube, as shown in figure 2.4. With the cone in this position, the diameter of the cone's base would equal the 6-centimeter side-length of the cube. Part 1 presents this choice and calls on students to compute the volume of the resulting cone. Since the volume of a cone is one-third of the volume of a cylinder with the same height and base, it is no surprise that the volume of the cone in figure 2.4 is 18π cm³, or one-third of the volume of the maximum cylinder in Chip off the Old Block.

What is surprising is that there exists a cone of significantly greater volume that the woodcarver could cut from the 6-centimeter cube. "A Base on a Face" hints at such a cone in the last question.

Students can investigate 2-D shapes that result when a plane intersects a polyhedron by using the applet Spinning and Slicing Polyhedra, which appears on the CD-ROM that accompanies this book.

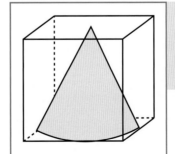

Fig. 2.4.

A cone cut from a cube, with the base of the cone on a face of the cube and the radius of the base equal to one-half of the side-length of the cube

Students who don't customarily think of 2-D shapes in a 3-D space might benefit from the supplement "Making a Model."

Part 2—"Going for the Max" (Optional)

A cone of maximum volume results from inscribing the cone's base in a regular hexagon formed by a plane that slices through the cube in a particular manner (see fig. 2.5). Part 2 of Cones from Cubes explores possible placements for this cone.

Chapter 2: Using Formulas to Measure Complex Shapes

Fig. **2.5.**

A cone with its base inscribed in a regular hexagon that is a cross section of a cube

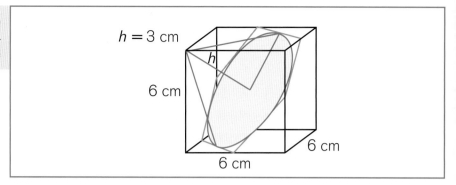

"Students should have frequent opportunities to formulate, grapple with, and solve complex problems ... and should then be encouraged to reflect on their thinking." (NCTM 2000, p. 52)

However, you may have some students who will need only the single clue at the end of "A Base on a Face" to launch on an independent search for such a cone. When these students have finished part 1, encourage them to find the dimensions of the maximum cone on their own, without the step-by-step guidance offered by part 2, "Going for the Max." In this form, the activity can become the powerful type of problem-solving experience that *Principles and Standards* recommends.

The applet Spinning and Slicing Polyhedra on the CD-ROM can help the students investigate the possibilities independently by allowing them to slice a cube with a plane and study the resulting cross section (see fig. 2.6). Although the applet is not essential to the activity, your students' use of it can generate many interesting questions and observations. This interactive tool can greatly facilitate students' explorations of the regions that can result when a plane slices a cube.

Another option for giving your students a hands-on experience with the cross sections is to have them slice cubes that they form with modeling dough or clay. The physical process of slicing and exposing cross sections can be very effective for some students.

You should anticipate that your students will come up with a variety of creative approaches to this problem. Many are likely to see the height of the new cone as half the length of the main diagonal of the cube. Some will find the radius of the cone as the apothem of the regular hexagon, using the equilateral triangle shown in figure 2.7.

Fig. **2.6.**

Images of (a) a cube and (b) a tetrahedron, sliced and opened to show the cross sections, from the applet Spinning and Slicing Polyhedra

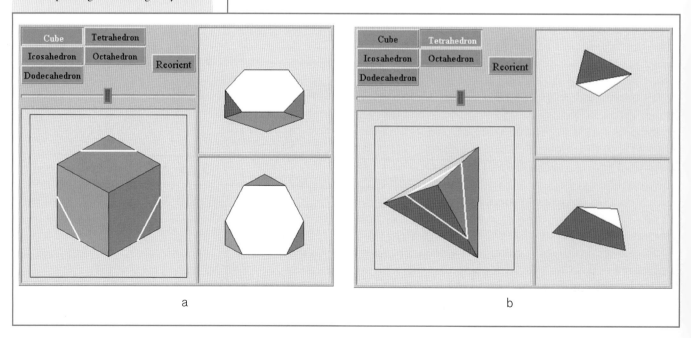

a

b

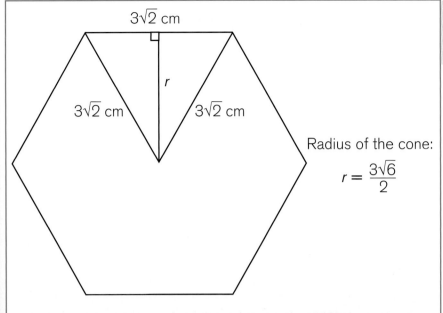

Fig. **2.7.**
A regular hexagon in which the base of a cone of maximum volume can be inscribed

Radius of the cone:
$$r = \frac{3\sqrt{6}}{2}$$

"*All students should ... apply appropriate techniques, tools and formulas to determine measurements.*"
(*NCTM 2000, p. 320*)

Other students will find the slant height of the cone by viewing the cube from the top. From this vantage point, they will note that the slant height must be three-fourths of the length of a diagonal on the top face of the cube (see fig. 2.8).

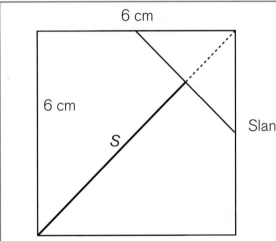

Slant height of the cone:
$$S = \frac{9\sqrt{2}}{2}$$

Fig. **2.8.**
The top face of the cube, with the top side of the hexagon showing

Of course, you may also have students who are unable to determine the dimensions of this new cone in the cube on their own. The optional part 2, "Going for the Max," will lead these students through the necessary steps in the computation of the volume without requiring them to develop the process entirely by themselves.

Even with the illustrations that "Going for the Max" provides, students will need substantial visualization skills to "see" the three-dimensional objects in the two-dimensional drawings of the cone in the cube. Some students can handle this problem entirely at an abstract level, visualizing the geometric relationships and making the corresponding measurements. Other students may need visual aids such as those that the applet Spinning and Slicing Polyhedra can provide.

Chapter 2: Using Formulas to Measure Complex Shapes

> *"All students should ... use visualization, spatial reasoning, and geometric modeling to solve problems."*
> (NCTM 2000, p. 308)

> *"All students should ... visualize three-dimensional objects from different perspectives and analyze their cross sections."*
> (NCTM 2000, p. 308)

Supplement—"Making a Model" (Optional)

You may have still other students who need to have the actual objects in their own hands before they can fully understand the measurement relation between the cone and the cube. These students may benefit from starting with a paper model. The supplement "Making a Model" guides students step by step in constructing a model that shows both the regular hexagon and the cone of maximum volume positioned in the cube. (See fig. 2.9.)

Seeing a regular hexagon on a cross section of a cube may surprise students. Even those who can see this hexagon clearly without a model often miss the fact that the hexagon cuts the cube symmetrically into two congruent halves. The model shows that a cone can be placed in each half, with its apex at a vertex of the cube.

This supplemental exercise can work well as a cooperative learning experience for students in small groups. The extra hands will be useful in constructing accurate models. Once the students complete their models, let them use metric rulers to measure and verify the dimensions that they computed theoretically in the earlier sections of the activity.

Sometimes, finding sufficient class time for students to make the models is difficult. As an alternative, you can make a model on a larger scale to use to illustrate this rather unusual orientation of a cone within a cube.

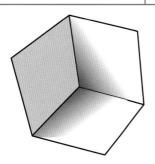 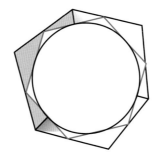 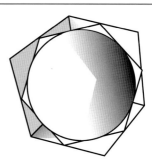

Fig. **2.9.**
Three stages in the construction of a model showing how to cut a cone of maximum volume from a cube

Assessment

Assess your students' facility with the formula for the volume of a cone. Be sure that they make appropriate use of relevant geometric terms, such as *slant height* and *apothem*. Also pay close attention to whether or not your students are able to visualize the situations described in the activity sheets. Use your observations to help you decide which individuals or classes would benefit from the particular visual aids and hands-on supplements offered on the CD-ROM.

It is essential for students to understand that slicing a cube with a plane can result in many cross sections that are not squares. They must also understand various ways to visualize the key geometric relationships that a solid has with the cube from which it is cut, and they should be able to represent these relationships in one, two, and three dimensions.

For a good performance assessment of your students' understanding of these ideas, you might ask them to undertake a task that is similar to the woodcarver's tasks in Chip off the Old Block and Cones from

Cubes. Have your students construct a model of the pyramid that a woodcarver could cut from a 6-centimeter cube, assuming that the pyramid's base is an equilateral triangle and that the pyramid's volume is the maximum that the woodcarver can obtain.

Where to Go Next in Instruction

Good problems often lend themselves to a variety of approaches. For example, a careful study of figure 2.5 reveals several different mathematical relationships among the segments drawn and the elements that they represent in the cube and cone. Thus, in determining the dimensions of the cones under consideration in Cones from Cubes, students have opportunities, as called for in *Principles and Standards*, to apply any of various mathematical arguments and techniques that they deem appropriate.

In the next activity, Measuring a Geometric Iteration, similar opportunities present themselves as students examine simple cases in an iterative process. They draw on various mathematical concepts and strategies as they look for patterns in the increasingly complex shapes that the iteration produces.

Measuring a Geometric Iteration

Goals

- Understand the nature of an iterative algorithm in geometry
- Visualize the results of a geometric iteration in a plane
- Compute, analyze, and generalize changing measurements

Materials and Equipment

For each student—
- A copy of the activity sheet "Iterating on a Plane"
- A ruler
- Scissors
- Several sheets of paper (different colors, if possible), a pencil, and a calculator
- A copy of the activity sheet "Iterating in 3-D" (optional as a comprehensive assessment)

pp. 112–14; 115–16

Discussion

In the middle grades, students customarily learn to compute perimeters and areas of specific geometric figures. The simplest of those figures is a square. Many students see nothing special about a square beyond the fact that its shape makes it convenient to use as a basic unit of area. However, if these same students repeatedly apply certain reconstruction processes to a "simple" square, they may be surprised by the complex structures that they can create. Measuring a Geometric Iteration demonstrates how a single iteration algorithm can generate a sequence of geometric figures that increase in complexity from stage to stage.

Part 1—"Iterating on a Plane"

Students begin by using a ruler to draw a 3-inch square, which they cut out. After tracing its outline in pencil on a clean sheet of paper, they cut their square into nine 1-inch squares. They throw away four of their smaller squares and arrange the remaining five in the square that they traced from the original 3-inch square. They place one 1-inch square in each of the four corners of the 3-inch square and the fifth square in the center (see fig. 2.10).

The students imagine repeating this process infinitely with each new, smaller square that the process produces as it continues from stage to

Fig. 2.10.

The first stages of an iterative pattern beginning with a square

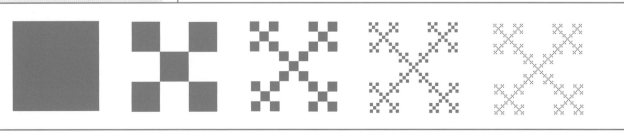

stage. They investigate the areas and perimeters of the shapes that result at early stages of the iterative process, and they look for patterns in the changes from one stage to the next. Used this way, measurement becomes a vehicle for examining and explaining new mathematical patterns and concepts that reach well beyond the familiar and the usual.

At the heart of this activity is the iteration rule, which the activity page states in words and illustrates in pictures. Some students may never have encountered an iteration algorithm with a geometric application. Therefore, it is important to take the time to discuss the step-by-step process in detail. You should emphasize that the square is the stage 0, or initial, figure only. It gives way to the stage 1 figure, but its outline forms a square frame into which all the subsequent stages fit. Students need to see clearly that the new figure at each successive stage—that is, the "input figure" for the iterative process—is the "output figure" of the previous stage.

The students enter values in a data table that shows the number of squares, the area, and the perimeter of the shape at stage 0, stage 1, stage 2, stage 3 and stage 4 of the process (see fig. 2.11). The students then search for any numerical and algebraic patterns in the data from these early stages, and later they try to generalize their results for the nth stage of the process.

Fig. 2.11.

Sample table showing data from the early stages of the iterative process as well as from stage n

Stage	0	1	2	3	4	n
Number of squares	1	5^1	5^2	5^3	5^4	5^n
Total area (in^2)	9	$9(5/9)$	$9(5/9)^2$	$9(5/9)^3$	$9(5/9)^4$	$9(5/9)^n$
Total perimeter (in^2)	12	$12(5/3)$	$12(5/3)^2$	$12(5/3)^3$	$12(5/3)^4$	$12(5/3)^n$

Students should note that the number of squares in the figure at any stage is always five times the number at the preceding stage. Thus, in the data on the number of squares, a diverging geometric series emerges, in the form of a sequence of successive powers of 5. When analyzing the areas and perimeters of the figures at each successive stage, students should encounter two very different geometric sequences. The successive areas form a geometric sequence that has a constant multiplier of 5/9 and converges to 0, and the successive perimeters form a geometric series that has a constant multiplier of 5/3 and diverges.

When your students are looking for patterns, remind them regularly of the importance of representation. For example, if students evaluate their results in decimal form when they enter values in the data table on the activity sheet, no patterns will catch their eye. However, if they express their area and perimeter calculations in fractions with exponents, the patterns will be immediately apparent.

Encountering a situation in which a geometric figure can have a tiny area and an astronomical perimeter will challenge the intuitions of many students. You may have some students who struggle hard to grasp such a notion. The last two questions of the activity will help students gain a better appreciation for what the divergence of the perimeter really means.

These questions help students understand the rapidly increasingly perimeter in terms of real-world measurements. First, they use a calculator to find the stage in the iterative process at which the total number of smallest squares exceeds 6.5 billion, often given as the total number of people in the world. They may be surprised to discover that the number of squares at stage 14 is about 6.1 billion, but stage 15, with 30.5 billion, is the first to exceed 6.5 billion. Students then find the stage at which the total perimeter will exceed one mile. They need wait only for stage 17 for this event.

Students may be astonished—and awed—when faced with the fact that, within the boundary of the initial 3-inch square, a stage 17 figure exists that is so complex that its outer boundary is greater than one mile in length. Indeed, students must accept the fact that a stage of the figure exists—still within the boundary of the 3-inch frame—that has a perimeter greater than any enormous length they may choose and that also has an area smaller than any minuscule area they may choose!

The figures that this iterative algorithm produces are truly beautiful, and the successive figures quickly become incredibly complex. They also raise the question of limits. As the stage number goes to infinity, the limiting figure can be described as a fractal. It is an abstract structure in 2 dimensions with no area. Yet the limit figure is made up entirely of connected line segments and, of course, points. In one sense, it is not very different from the stage 4 image shown in figure 2.10. The limit figure—the fractal—is infinitely complex, with every part, however small, containing an exact, reduced version of the entire structure.

Assessment

You can assess your students' understanding of the iteration rule by having them state it in their own words. Focus on these three parts: *reduction*, *replication*, and *rebuilding*. For example, the following description covers the three elements: "Make 5 copies at a scale of 1/3, and place them in the corners and center of the frame."

Also assess your students' skills in using fractions and exponents in representing the values of the area and perimeter. Don't overlook their ability to generalize the process in terms of the variable *n*. Moreover, it is important to assess their understanding of the measurements, the geometry, and the notions of limits that the activity involves, at a deeper level. You can use the following questions to test their understanding:

- "If the limit of the converging areas is 0, is there a limit figure?"
- "If so, what is the area of the limit figure, and how can the area be described?"

Part 2—"Iterating in 3-D" (Optional Assessment)

When your students have completed "Iterating on a Plane," you can obtain a comprehensive assessment of their understanding by posing an analogous problem in 3-dimensional space. The activity page "Iterating in 3-D" gives a sample of such an assessment.

Here students start with a 3-inch cube. They use a scale of 1/3 to make 9 copies of the cube, and they place them at the corners and center of the frame of the original cube. Figure 2.12 shows the iterative process.

Before assigning this problem to the students, discuss with them the iteration rule that it gives for generating successive figures. In this problem, the number of cubes in any stage is always nine times the

"All students should ... apply informal concepts of successive approximation ... and limits in measurement situations."
(NCTM 2000, p. 320)

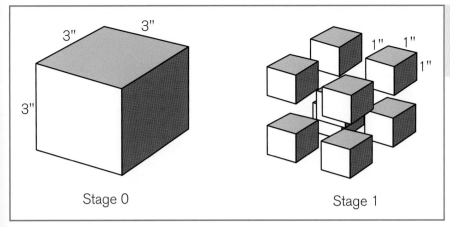

Fig. **2.11.**
Stages 0 and 1 of a 3-D iterative pattern beginning with a cube

number at the preceding stage. The successive volumes produce a geometric sequence that has a multiplier of 1/3 and converges to 0, but the successive surface areas, as many will be surprised to discover, remain constant at 54 square inches.

Conclusion

The activities in this chapter have illustrated the use of basic formulas for area and volume to explore increasingly difficult measurement problems in theoretical settings. In addition to gaining an appreciation for the power of these formulas, students should have acquired some understanding of how mathematicians discovered and developed these formulas.

Although mathematical formulas can often produce exact answers to measurement questions that arise in theoretical settings, not all formulas are so exact. This is an important point for students to understand. Chapter 3 demonstrates some ways in which students can develop these insights.

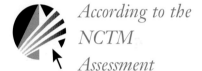

According to the NCTM Assessment Principle, "it is important that assessment tasks be worthy of students' time and attention. Activities that are consistent with (and sometimes the same as) the activities used in instruction should be included."
(NCTM 2000, p. 22)

NAVIGATING through MEASUREMENT

GRADES 9–12

Chapter 3
Discovering and Creating Measurement Formulas

A measurement formula describes a procedure for calculating one measurement by using related measurements. Ideally, the formula provides a shortcut for deriving the measurement. For example, instead of counting squares to find the area of a rectangle, students learn to measure the rectangle's length and width, and then they apply the formula $A = lw$. But do students understand where formulas come from? *Principles and Standards for School Mathematics* (NCTM 2000) urges us to help our students make sense of formulas and realize that if someone asks, "But where does the formula come from?" it's not acceptable to respond, "It's the law!"

Students use a formula more effectively after discovering and developing it for themselves. The student who has sliced a rectangle or a parallelogram in half along a diagonal has a better understanding of why the formula for the area of a triangle is $A = 1/2(bh)$. Formulas for volume, like their counterparts for measurements in two dimensions, are also interrelated. Just as a triangle has 1/2 the area of a corresponding parallelogram, a pyramid has 1/3 the volume of a corresponding prism.

The first activity in this chapter, Discovering the Volume of a Pyramid, allows students to uncover the relationship between the volumes of a pyramid and a prism. This activity helps them develop the corresponding formulas on their own.

Mathematical measurements use precise, provable formulas that yield exact measurements of mathematical quantities. Beyond the realm of these strict formulas lies another realm of formulas, encompassing the less absolute, less mathematically precise formulas that we use to quantify information that is nonmathematical or not completely mathematical. Though the mathematics of these formulas is not as pure as that of the formulas in the first realm, the measurements that they yield influence decisions that profoundly affect our lives.

How does someone measure the quality of a school, the success of a product, the rate of unemployment, the value of a power resource, the popularity of a leader, or even the size of a tree? Students should learn that formulas for such measurements are somewhat arbitrary; they reflect the purposes, needs, tastes, and biases of the people who create them.

The second activity in this chapter, Measuring the Size of a Tree, gives students a firsthand experience of the decision making involved in developing a formula that is not entirely based on mathematical concepts. The activity invites the students to participate in the process of creating one of these formulas for themselves.

You can have your students work individually on the blackline pages for both activities in the chapter, but each student does not need to have all the other materials. You can ask small groups of students to share materials such as metersticks, spreadsheet software, and sets of wooden cubes. The second activity, Measuring the Size of a Tree, can also be highly effective if students work in groups from start to finish. You may in fact decide that this is the best way for your students to do the activity.

Discovering the Volume of a Pyramid

Goals

- Discover the formula for measuring the volume of a pyramid
- Understand that a shape with smooth sides can be measured by successive approximations with shapes with jagged sides
- Use technology to explore the limit of a process involving the ratio of two measurements

Materials and Equipment

For each student—
- A copy of the activity sheet "Building Pyramids with Cubes"
- A copy of the activity sheet "Probing Pyramids with Spreadsheets"
- A pencil

For each group of three or four students—
- Approximately 40 small wooden blocks of equal size
- Access to spreadsheet software

pp. 117–18; 119–20

Discussion

The activity Discovering the Volume of a Pyramid has two parts. "Building Pyramids with Cubes" engages students in a hands-on construction of "pyramids" made with cubes (see fig. 3.1), and "Probing Pyramids with Spreadsheets" invites students to use an electronic spreadsheet to extend their exploration of these cube-pyramids.

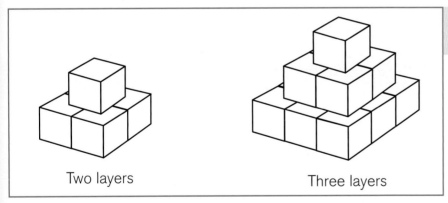

Two layers   Three layers

Fig. **3.1.**
Cube "pyramids" with square bases

The two parts of this investigation supply empirical evidence of the constant 1/3, which figures in the formula for the volume of a pyramid. Ultimately, the students will put together all their information about the volume (V) of a pyramid to understand that

$$V = \frac{1}{3} Bh,$$

where B stands for the area of the base of the pyramid and h stands for its height.

Chapter 3: Discovering and Creating Measurement Formulas

Part 1—"Building Pyramids with Cubes"

Students begin their exploration by building approximations of right square pyramids with cubes. Students can work in small groups to construct the cube-pyramids while completing their own activity sheets. Figure 3.2 shows the process of stacking square layers of cubes one on top of the other, with each higher layer having one fewer cubes on a side than the layer below it.

Fig. **3.2.**

Students construct cube "pyramids" and explore their volumes

Because the students can easily calculate the volume of a "pyramid" made with cubes, they can readily compare such a pyramid's volume with the volume of a prism that has the same base and height. As they measure the volumes of corresponding cube-pyramids and prisms, they enter the values in a data table. (The activity assumes that students already understand the formula for the volume of a prism.)

As the cube-pyramids get larger through the addition of layers, their jagged edges become less significant in relation to their total size. Thus, the shapes become better and better approximations of actual, smooth-sided pyramids. As this process of producing ever-larger pyramids and comparing them to corresponding prisms continues, the ratio between the volumes of the cube-pyramids and the volumes of the related prisms converges to 1/3. Figure 3.3 shows a graph of 200 stages of the process.

However, your students' work in part 1 of the activity will not necessarily lead them to the conclusion that the ratio of the volumes in fact converges to 1/3. This result may become clear to the students only through their work in part 2, when they collect much more data with the help of an electronic spreadsheet.

As students work through part 1, they may soon tire of building the solids from blocks and may simply fill in the table by using the numerical patterns that they see. This movement away from concrete building shows progress toward an abstract understanding. Students will readily observe the pattern for the simple cubic prisms: the volume of a prism is the cube of the prism's side length. For the volumes of the pyramids

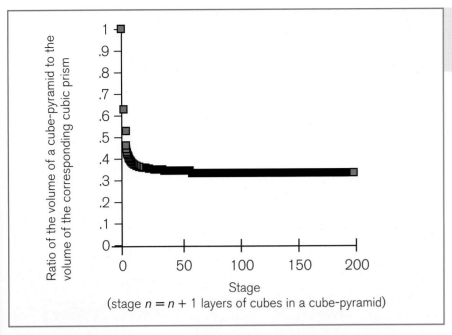

Fig. **3.3.**
Spreadsheet graph of the first 200 stages of the process

of cubes, students are likely to observe that if they square the side length of a new cube-pyramid's base and then add this product to the volume of the previous cube-pyramid, they can find the volume of each new cube-pyramid.

This recursive formula represents an important pattern, and it will work well on a spreadsheet when the students get to part 2 of the activity. However, students' attempts in part 1 to use the pattern for the volumes of the cube-pyramids to write an explicit formula for the volume of a cube-pyramid of n layers may prove unsuccessful. Depending on the level of your class, you may want to make this more challenging task an optional opportunity for extra credit.

Activities in chapter 2 illustrated contexts in which representing data with fractions facilitated the discovery of patterns and trends. By contrast, Discovering the Volume of a Pyramid provides an example of a situation in which a decimal representation works best.

As the students work in part 1, they will see that if they convert the fractions in the ratio column in their data table to decimals, they will more easily see that the values slowly decrease. However, the convergence is so gradual that it is almost impossible to observe without extending the table for tens—or even hundreds—of rows.

In fact, most students will predict—incorrectly, as it happens—that the ratio will eventually converge to zero. Using spreadsheet technology to gather more data can correct this misunderstanding.

Part 2—"Probing Pyramids with Spreadsheets"

In the second part of the activity, students continue to compare the volumes of increasingly large cube-pyramids to the volumes of corresponding prisms, but now they harness the power of electronic spreadsheets to extend their investigations. The activity sheet "Probing Pyramids with Spreadsheets" gives the students step-by-step instructions for building their spreadsheets.

Many students are quite proficient with spreadsheet software and will not need these instructions. If you think your students can build the spreadsheet on their own, you can have them work directly from the

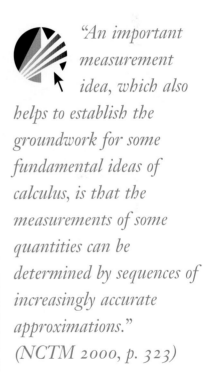

"An important measurement idea, which also helps to establish the groundwork for some fundamental ideas of calculus, is that the measurements of some quantities can be determined by sequences of increasingly accurate approximations."
(NCTM 2000, p. 323)

data table in "Building Pyramids with Cubes" instead of giving them the second activity sheet, "Probing Pyramids with Spreadsheets."

Students will be most convinced of the convergence to 1/3 when they "see" it on a graph. The spreadsheet graph provides a dramatic demonstration. The convergence in the decimal numbers in the ratio column on the spreadsheet is often less persuasive than the evident convergence of the values on a spreadsheet graph (see fig. 3.3).

Assessment

You might be surprised to learn that many students who discover the convergence to $.\overline{3}$ nonetheless fail to realize that this represents the ratio of 1:3. Thus, it is often appropriate to merge this lesson with an old-fashioned measurement demonstration using actual plastic models of a prism and pyramid. Make sure that your two solids have congruent bases and heights. Then have your students predict how many times you can fill the pyramid with water or rice and pour the contents into the prism before the prism is full.

If your students predict that only two "pyramids' worth" will completely fill the prism, they are not making the connection between their spreadsheet results and the physical models. To emphasize the 1 to 3 ratio that is the point of this activity, have a student carefully fill the pyramid model with water or rice and pour the contents into the prism while the rest of the students in the class count to three.

Because all stages of the activity use blocks of the same size, the cube-pyramids grow in size very quickly. It is important that students not conclude that the volume formula that they have discovered for a pyramid is accurate only for an extremely large pyramid or for a pyramid whose height equals the length of its base.

You might ask your students to write entries in their mathematics journals explaining how a "pyramid" of cubes can closely approximate the volume of a conventional, smooth-sided pyramid whose height is equal to the length of its base. Their journal entries can extend this idea to pyramids whose heights and base lengths are different.

For a final assessment, you might have your students use their newfound formula to estimate the volume of an actual pyramid, such as one of the ancient pyramids of Egypt.

Where to Go Next in Instruction

The formula for the volume of a pyramid is an exact mathematical formula. Because *volume* and *pyramid* are attributes with exact mathematical definitions, we can prove this formula mathematically. Not all formulas are so neat and tidy. In the following activity, students explore a context in which the objects and attributes of interest do not have purely mathematical definitions, and formulas for measuring them are neither exact nor mathematically verifiable.

Measuring the Size of a Tree

Goals

- Define and develop a formula for a measurement in a real context
- Experiment with a variety of measurement techniques, including estimation and indirect measurement
- Explore the variety of measurable attributes and their relevance to everyday objects
- Use variables to represent unknown measurements and show their relationships in formulas for another measurement

Materials and Equipment

For each student—
- A copy of the activity sheet "Making a Formula"
- A copy of the activity sheet "Using Your Formula"
- A copy of the activity sheet "Reporting Your Results"
- A copy of the activity sheet "Assessing a Poster" (optional)
- Clean sheets of paper or a poster board (for each group if your students are working in groups instead of individually)

For each group of four or five students—
- Basic measuring tools, such as metersticks, measuring tapes, and calculators

pp. 121–22; 123–25; 126; 127–28

Discussion

The activity Measuring the Size of a Tree gives students an opportunity to devise a measurement formula on their own. This activity unfolds in three parts, with an optional fourth part. Students move from the preliminaries of looking at trees to the development of a formula ("Making a Formula"), the testing of the formula ("Using Your Formula"), and the presentation of results and reflections ("Reporting Your Results"). If you choose, you can then have your students engage in a peer-assessment of a poster ("Assessing a Poster").

You can have your students work individually, sharing only materials, or you can have them work together in groups from start to finish. Students working in groups can share ideas and strategies as well as materials, and they can devise and test a formula collaboratively.

Working in groups may well increase your students' learning in the activity. The biggest adventure for students in this activity is developing an original formula for calculating size. Going outdoors and experiencing mathematical measurement beyond the classroom can heighten the element of adventure, and the process of making difficult decisions in a cooperative setting can enhance the students' feelings of creativity, autonomy, and invention. Many students report that learning to work together is one of the biggest educational benefits of this activity.

In assessing the project, one girl said, "The best part was choosing our own formula."

Part 1—"Making a Formula"

Students begin by going outside and studying a tree. Their first task is to list twenty attributes that they could measure for their tree—its number of leaves and the maximum area of shade that it creates, for example. Having students come up with so many different attributes encourages them to think beyond the standard measurements of height and circumference. It prompts them to wonder about characteristics like greenness of leaves, number of dead branches, age, weight, number of insects living in the tree, and so on.

The stage is now set for students to undertake the task at the heart of the activity—devising an original formula for measuring the size of a tree. Focusing specifically on attributes that relate to the size of the tree, the students choose three or four such attributes whose measures they think would be very important for determining the overall size of a tree. They consider how to express these attributes as variables and how to set them in a formula.

The students must decide how important they think each of their attributes is in relation to the others and give each one appropriate "weight" in their formulas. They evaluate the strengths and weaknesses of different possibilities and grapple with issues of reliability and validity as they develop their formulas, write them, and explain them. The solutions provided in the appendix include some actual responses from ninth graders to these open-ended challenges.

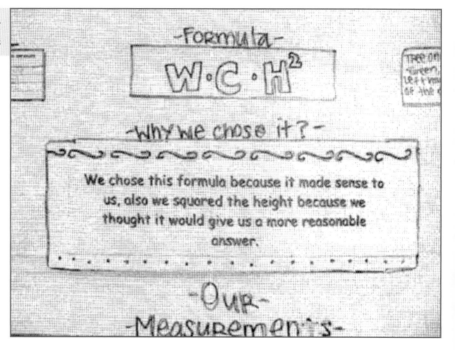

Fig. **3.4.**
A formula developed by students

An elementary school class asked the U.S. Forest Service, "How many recycled newspapers does it take to save a tree?" To answer the question, a Forest Service ecologist determined how many newspapers of average weight could come from an average tree. He used standard formulas to do this, calculating the weight of usable pulp from an average Douglas fir tree (90 feet tall with a 12-inch diameter at breast height and grown for harvest as timber). On the basis of his results, he concluded, "If the families of all 30 students [in the class] recycle, then you're saving about 15 trees per year." For details, see www.fs.fed.us/pnw/kids/recycled.shtml.

If your students are working in groups, each group may develop a different formula, depending on the agreements that the group members make about how to measure the size of a tree. Many formulas are possible, of course, and students may come up with a surprising array. Indeed, even the U.S. Forest Service uses various and sometimes proprietary formulas for estimating the size of a stand of trees in board feet. The students' formulas in this activity are far from secret. Each student or group of students must defend the formula in a final presentation to the class.

Part 2—"Using Your Formula"

In part 2 of the activity, students take their formulas beyond the development stage, exploring how well the formulas work by using them to compare one tree to another. The students have already decided which measurable attributes of a tree to enter into their formulas for the size of a tree. Now they must determine how to quantify these attributes.

To make their measurements, students may need to develop other measurement tools in addition to the standard tools that you have provided. For example, to measure the height of a tree, students might need to measure the angle of inclination from a level line of sight to the top of the tree. Some might simply use the fist method described in chapter 1. Others might create a simple tool with a protractor, a plumb line (a string and a weight), and rolled paper or a straw. (See fig. 3.5.) Be prepared to help students select and obtain appropriate materials.

Fig. **3.5.**
Students measuring an angle of inclination

Students should discover that the units they use can greatly affect the numbers that result from calculations with their formulas. For example, a height measured in inches can have 12 times as much influence in a calculation as the same height measured in feet. Students should also realize that in many cases the units they end up with when they apply their formula will not be familiar ones such as cubic inches or square feet. Often, the units will turn out to be something very complicated. Students usually enjoy inventing names for these unorthodox units.

The U.S. Forest Service is interested in the size of a tree mainly for calculating the amount of timber in it. The quantity of timber can give a measure of the tree's value—its "worth" in market terms. By contrast, your students might be interested in counting the leaves or branches of

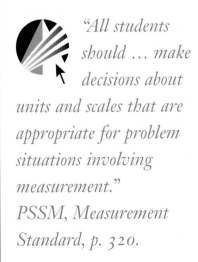

"All students should ... make decisions about units and scales that are appropriate for problem situations involving measurement."
PSSM, Measurement Standard, p. 320.

Chapter 3: Discovering and Creating Measurement Formulas

their trees without any intention of measuring the tree's value, commercial or otherwise. This difference can demonstrate to the students that a formula for the size of a tree truly connects with issues of politics, use, need, and taste.

The sample student formula in the solutions section uses variables for height, girth (or circumference), number of branches, and average length of each leaf. Initially, some students even selected the weight of a tree as an attribute to use in their formulas, but they ultimately abandoned it when they realized how difficult it would be to measure. Figure 3.6 shows how students in one group described their method of measuring a tree's height.

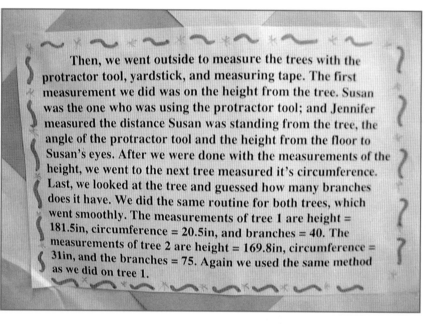

Fig. **3.6.**

Students' account of their method for measuring the height of a tree

Before your students actually make measurements and apply their formulas, the activity asks them to estimate the relative sizes of two trees and record their guesses. They complete these tasks by filling in the blank and circling the appropriate word in the following sentence: "Tree 1 is _____ times bigger / smaller than tree 2." The students' estimates can provide them with a check on the reliability and validity of their formulas. Suppose that they estimate tree 1 as bigger than tree 2 before they measure and apply their formula, but after measuring and applying the formula, they find themselves looking at results that show tree 2 as bigger than tree 1. This discrepancy could represent a problem in their formula or measuring techniques.

Making and using their own formulas can lead students to a wide variety of topics in the mathematics curriculum. For example, the first-semester ninth graders pictured in figure 3.5 used a protractor tool that prompted them to learn about the tangent ratio, a concept that ninth graders do not typically encounter. (Fig. 3.7 shows the students' method for using the tangent ratio to find the height of a tree.) Also, students who choose to use the tree's diameter as one of the measurements in their formula will need to estimate this measurement from the tree's circumference. For many of them, this will be the first time that they have a "real-world" use for the number π.

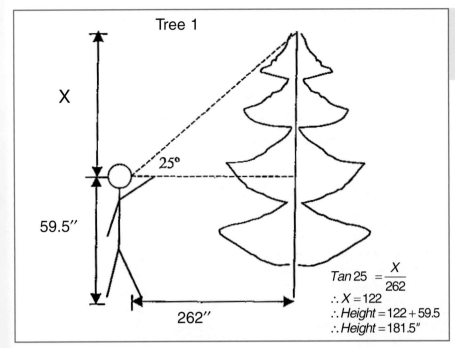

Fig. **3.7.**

Students' diagram (not to scale), showing their use of the tangent ratio in finding the height of a tree

Assessment

For a final assessment of your students' work, you can ask the students to report their results and reflect on them. Students can organize their results on clean sheets of paper or a poster board.

Part 3—"Reporting Your Results"

The activity sheet "Reporting Your Results" gives students specific instructions for organizing their ideas and presenting them to you or the entire class for a final assessment. Students are to present and explain their formulas and summarize their work by answering the following questions:

- "Looking over your results, would you change your formula at all?"
- "What did you learn from this project?"
- "What was your favorite part of this project?"
- "What was the most difficult part of the project?"

Students are also directed to thank classmates who helped them (explaining the roles of all group members if they worked in a group), and they are asked to acknowledge outside sources of information. Figure 3.8 shows one group's responses.

Writing up their work and talking about it can cement the students' sense of accomplishment. Students' presentations of their posters to the class can consolidate their learning from the project.

Part 4—"Assessing a Poster" (Optional)

The activity sheet "Assessing a Poster" provides a rubric for evaluating students' presentations of their work. If you choose, you can distribute copies of this rubric to your students and assign them the task of assessing the work of particular peers as they present their results to the class. The rubric calls for assessment of the following aspects of the report:

Fig. **3.8.**
A group's summary on a poster

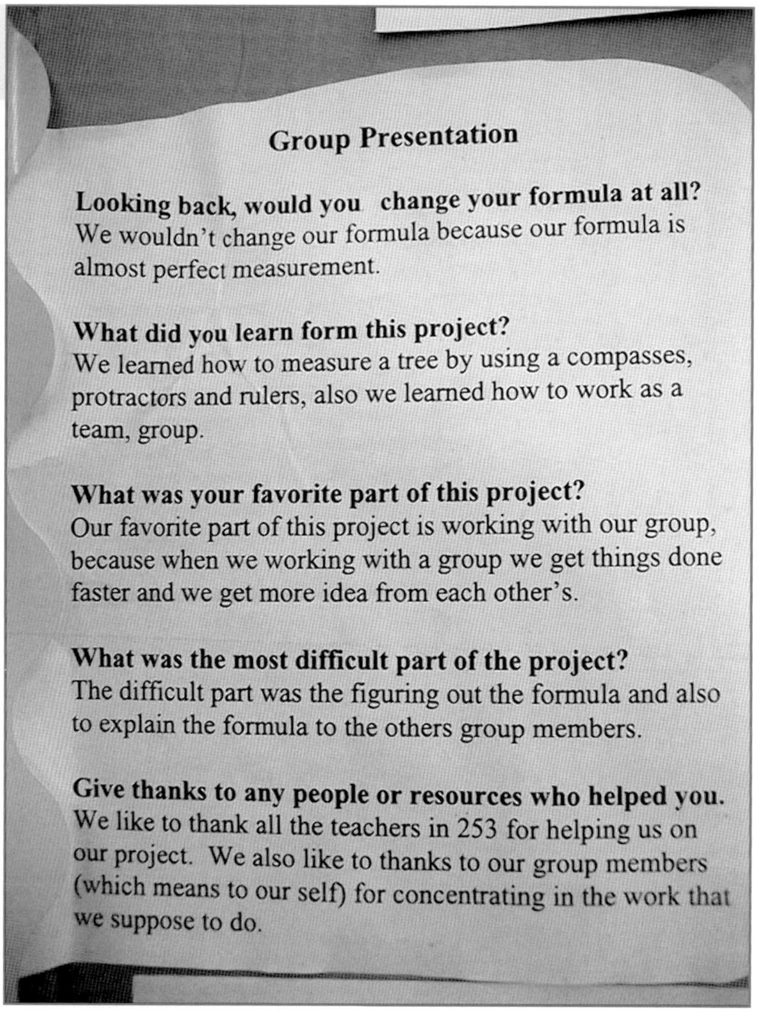

> **Group Presentation**
>
> **Looking back, would you change your formula at all?**
> We wouldn't change our formula because our formula is almost perfect measurement.
>
> **What did you learn form this project?**
> We learned how to measure a tree by using a compasses, protractors and rulers, also we learned how to work as a team, group.
>
> **What was your favorite part of this project?**
> Our favorite part of this project is working with our group, because when we working with a group we get things done faster and we get more idea from each other's.
>
> **What was the most difficult part of the project?**
> The difficult part was the figuring out the formula and also to explain the formula to the others group members.
>
> **Give thanks to any people or resources who helped you.**
> We like to thank all the teachers in 253 for helping us on our project. We also like to thanks to our group members (which means to our self) for concentrating in the work that we suppose to do.

- completeness
- mathematical content
- organization, layout, and creativity
- oral presentation.

The activity sheet specifies possible point values for students' work in each of these areas, with a total of 100 points for students who earn the maximum number of points in each category. These point values are of course arbitrary to some extent. Feel free to change them to suit the emphases in your teaching if you wish. You may want to discuss the values with your students and invite their input.

Having your students work with the rubric and use it to assess the work of peers can extend their learning from the activity in a very direct way. Thinking about and manipulating an assessment rubric can reinforce ideas about formulas that are not strictly or precisely mathematical but provide valuable tools for measurement in actual situations. The rubric shows another measurement formula—an assessment formula—that depends on the selection of variables and the assignment of values and weights to make a useful and meaningful measurement.

However, you may prefer not to have your students assess peers but instead reserve the blackline page for your own use in assessing your students' work on the project. In this case, you may want to add another category—individual participation—to those specified on the activity sheet. This new category will give you an opportunity to assess whether or not your students made productive use of the class time that you set aside for the project. You might want to make the students' efforts in this category worth as much as 20 points or more in addition to the 100 possible points on the rubric.

When you or your students are assessing the mathematical content of the work, be sure to give students' use of units particular attention. Some students have difficulty coming up with appropriate units for their measurements of the size of the tree. The poster presentations will provide good opportunities to discuss each case that arises.

Conclusion

This chapter has focused on the development of measurement formulas. Formulas are mathematical models—but not our only mathematical models—for deriving measurements. The next chapter explores how students can use other mathematical models besides formulas to derive some historically important measurements. The final chapter of the book extends this investigation, examining ways of letting students see how technology-based measurements can assist in developing a wide range of mathematical models for real-world measurement.

You can add another category to "Assessing a Poster" to help you assess your students' work:

INDIVIDUAL PARTICIPATION
The student worked hard the entire time and fulfilled his or her responsibilities.

Grade_____ out of 20 points

NAVIGATIONS SERIES

GRADES 9–12

NAVIGATING *through* MEASUREMENT

Chapter 4
Classic Examples of Measurement

The lower part of a cone or pyramid cut by a plane parallel to the base is the frustum.

Frustum

"In addition to reading measurements directly from instruments, students should have calculated distances indirectly and used derived measures."

(NCTM 2000, p. 321)

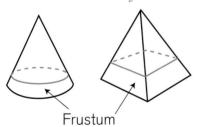

Throughout history, scientists and mathematicians have derived formulas and procedures that have built on simpler, easier measures to achieve measures that are more difficult or complex. Historians still wonder how the ancient Egyptians developed a formula for the volume of the frustum of a pyramid—a formula whose rigorous proof requires limit arguments akin to those at the heart of calculus. Whether the stories about Thales (ca. 640–540 B.C.) are myth or fact, they celebrate many brilliant measurement feats, including the use of shadows to measure the heights of pyramids and the application of proportional reasoning associated with similar triangles to measure the distance from shore to ships at sea.

The list of inspired uses of relatively simple mathematics to make other, more complicated measurements—sometimes with a surprising degree of accuracy—is impressive. This chapter selects examples from the list for you to share with your students, in keeping with the recommendation of *Principles and Standards for School Mathematics* (NCTM 2000) that teachers help their students experience the power of mathematics through indirect measurement.

The activities in this chapter explore intriguing settings from the history of mathematics and show how these contexts can engage high school students in the process of measuring and the analysis of measurement error. Three activities highlight some ancient methods and mathematical models for measuring sizes and distances pertaining to the earth, moon, and sun. These methods and models are early, truly remarkable examples of the use of simple observations and basic mathematics to make challenging measurements.

Standard textbook problems pertaining to measurement can often lead to productive hands-on investigations, which in turn can deepen students' understanding of the mathematics in the problems.

The first activity, If the Earth Is Round, How Big Is It? takes students through the steps of measuring the circumference and diameter of the earth by a method that Eratosthenes invented in the third century B.C. Many high school geometry textbooks illustrate Eratosthenes' method in notes or exercises. Students who have studied basic right triangle trigonometry can use their knowledge to facilitate their work, but the exploration does not depend on trigonometry.

The second activity, Moon Ratios, follows up this investigation of ancient measurements of the earth with an exploration of methods that early astronomers developed for finding the visual angle of the moon and the distance from the earth to the moon. Students explore methods that enabled early astronomers to measure important ratios—the ratio of the earth-moon distance to the moon's diameter and the ratio of the diameter of the moon to the diameter of the earth. An acquaintance with right triangle trigonometry can expedite students' work in the activity but is not essential to it.

The third activity, How Far Is the Sun? lets students put elements of the strategies from the first two activities together to make estimates of the distance of the sun from the earth. This activity depends on a basic trigonometric fact about right triangles—that the ratio of either acute angle's adjacent side to the triangle's hypotenuse is equal to the cosine of the angle. Thus, students who are acquainted with basic right triangle trigonometry will understand the mathematics that is involved. However, if teachers supply relevant information, students can complete the activity without a grounding in trigonometry.

If the Earth Is Round, How Big Is It?

Goals

- Simulate a classical measurement process
- Analyze measurement error resulting from the measurement instrument

Materials and Equipment

For each student—
- A copy of the activity sheet "If the Earth Is Round, How Big Is It?"

For each group of three or four students—
- One or more Styrofoam balls (4–6 inches in diameter)
- Two straight pins (1.25–2 inches in length)
- A flashlight or pen light. (The class can share a movable bright light, if necessary.)
- A short tape measure (calibrated to millimeters; template provided)
- A 4-by-6-inch index card, scissors, and clear tape

For students who have not studied trigonometry—
- Centimeter grid paper (template provided)
- A protractor

For the class—
- A road atlas of the United States that students can use to find the distance (as the crow flies) from Bozeman, Montana, to Tucson, Arizona

pp. 129–30

The template "Measuring Tape" on the CD-ROM enables you to print and cut out short paper tape measures, calibrated to millimeters, for your students' use.

You can print grid paper for this activity from the template "Centimeter Grid Paper" on the CD-ROM.

Discussion

In the third century B.C., the Greek mathematician Eratosthenes devised a method for measuring the circumference of the earth. He measured the angle of the shadow cast by a vertical stick in Alexandria at noon on the summer solstice. The measure that he found for this angle (shown as $\angle TBA$ in fig. 4.1) was approximately 7.2 degrees. (The system of degrees that we use today to measure angles came into being after the time of Eratosthenes, but we can use degrees in carrying out his reasoning and method.)

Eratosthenes happened to know that on the same day, at the same time, due south in the town of Syene (S in fig. 4.1), the sun was directly overhead, and a vertical stick would cast no shadow. He also knew that the distance from Alexandria to Syene was approximately 5000 stades, or about 500 miles. Eratosthenes assumed that the sun's rays were parallel; hence, $\overline{TB} \parallel \overline{OS}$ in figure 4.1. Thus, $m\angle ABT = m\angle BOS \approx 7.2°$ Since 7.2° is equal to 1/50 of a circle, Eratosthenes concluded that the earth's circumference was 50×5000 stades, or 250,000 stades.

The size of a stade has been a subject of debate. One possible value, 559 feet to a stade (Eves 1990), would make Eratosthenes' estimate of the earth's circumference 139,750,000 feet, or 26,468 miles—a value

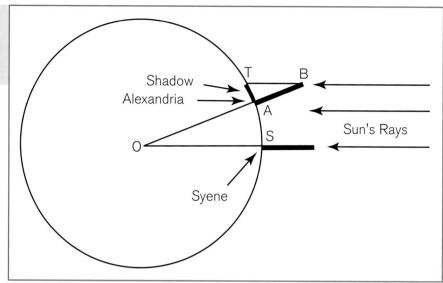

Fig. **4.1.**

The geometry of Eratosthenes' measurement of the earth's circumference (not to scale)

that is within a few percentage points of 25,000 miles, often given today as the earth's circumference. Using 559 feet per stade would make Eratosthenes' estimate of the earth's diameter 8,425 miles as compared with the contemporary measurement of the earth's polar diameter as 7,912 miles. The accuracy of Eratosthenes' estimates is rather amazing, considering the simplicity of his method and the absence of accurate instrumentation and sophisticated measurement systems in the third century B.C.

Like many geometry textbooks, the activity sheet presents a historical note explaining how Erastothenes (ca. 250 B.C.) measured the circumference and diameter of the earth. Using Erastothenes' ideas and data as starting points, students replicate his method and obtain his measurements in step 1 of the activity.

In step 2, students transfer Eratosthenes' method to the task of measuring the circumference and diameter of a ball. Equipped with a Styrofoam ball, a short paper tape measure, two straight pins, a bright light, an index card, scissors, and tape (and perhaps some grid paper and a protractor), students design a simulation of Eratosthenes' method. Their simulation should lead them to an accurate estimate of the diameter of the ball.

By asking students to design their own simulations, the exploration forces the students to think more deeply about the questions that the measurements raise than they would be likely to if they had a step-by-step procedure to follow. Students can be inventive in fashioning the materials to suit their simulations. For example, they might cut a strip from the index card and attach their paper measuring tape to it, stiffening and straightening the tape, as in figure 4.2.

What do the students gain by this? In the simulation pictured in figure 4.2, the measuring tape's straight segment allows the students to obtain a direct measurement of the shadow. By positioning their light source (not shown) so that its rays strike the top of the left-hand pin directly, the students allow the shadow from the right-hand pin to fall onto the straight segment of the ruler. The students can then inspect the right triangle whose legs are formed by the right-hand pin and its shadow. If the students happen to know right triangle trigonometry,

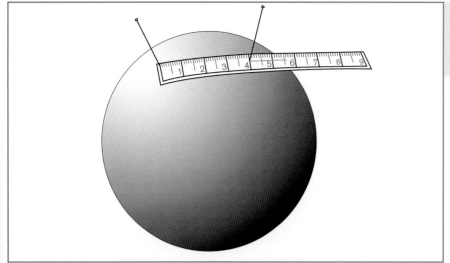

Fig. **4.2.**

A measuring tape extending on a tangent from a pin on a Styrofoam ball in a student simulation

they can compute the angle at the top of the triangle as the arctangent of the ratio of the shadow's length to the pin's height.

Students who have not studied trigonometry might also profitably set up their simulation as pictured in figure 4.2. They could use grid paper (or software for geometric drawing) to create a right triangle congruent to the one whose legs are the pin and its shadow, and then they could measure the angle directly with a protractor.

The activity challenges students to identify sources of error as well as to discuss alternative approaches. For example, if the light source is too close to the ball—say, within five feet—it will not be at all reasonable to assume that its rays are parallel, and any direct measurements that the students make can greatly distort their final, indirect measurement.

As students work with their simulations and make their measurements, they must also focus on the error that arises in working with an instrument (the paper measuring tape) with a particular degree of precision. The ruler provided for the activity allows students to make measurements that are precise to the nearest millimeter—that is, with a round-off error of less than 0.5 mm. This potential error has implications for all the measurements that the students derive from those that they make with the ruler. The solutions in the appendix include a sample showing how students might work out these implications.

Step 3 of the activity extends the students' investigation of Eratosthenes' method by asking them to consider a real-world case in which vertical sticks in two locations *both* cast shadows. Using a road atlas together with the information that Bozeman, Montana, is due north of Tucson, Arizona, students figure out how they could use simultaneous measurements of shadows in the two cities to estimate the circumference of the earth. The situation is sketched in figure 4.3.

Students must find a relationship between the central angle, labeled as ∠Z in figure 4.3, and the angles of the shadows, shown as ∠A and ∠D. Using the Exterior Angle Theorem, students can deduce that $m\angle Z = m\angle A - m\angle D$.

As an alternative approach to the two-shadow situation in step 3, you might prefer to have your students collaborate with those in another mathematics class in a city located on the same longitude line as your city but hundreds of miles away. The students in the two classes could then do the experiment "live" with telephone connections.

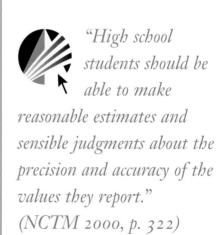

"High school students should be able to make reasonable estimates and sensible judgments about the precision and accuracy of the values they report."
(NCTM 2000, p. 322)

Fig. **4.3.**

A variation on Eratosthenes' method of measuring the earth's circumference, with vertical metersticks in two locations casting shadows.

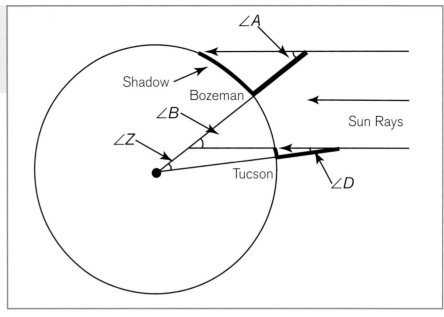

Because $m\angle A = m\angle B$,
$m\angle Z = m\angle A - m\angle D$.

Eratosthenes' effort to measure the circumference of the earth counters the myth that everyone thought the earth was flat until the age of Christopher Columbus.

Assessment

During the activity, be sure to assess your students' abilities to use the properties of circles and parallel lines to derive angles and arc lengths. You should also check their computation of the uncertainty intervals for their estimates of the circumference of the ball.

Students should gain two important insights from their work:

(1) Indirect measurements of a phenomenon typically depend on the development of a mathematical model, such as those shown in figures 4.1 and 4.3.

(2) These mathematical models usually depend on assumptions about the phenomenon, such as that the earth is a sphere and that the sun's rays are parallel.

To assess your students' understanding of these essential ideas, you might ask them to write journal entries evaluating the importance of the assumptions that they made in the simulation with the Styrofoam ball in step 2.

Encourage your students to reflect on what would happen to the shadows if various features of the situation were different. For example, how would a light that was very close to a ball affect the shadows of pins in the ball?

Where to Go Next in Instruction

In the history of mathematics, the measurement of the ratio of two quantities was sometimes just as important as, and often a prelude to, the measurement of a specific quantity. One very important ratio that students encounter is π. To the ancients, π was not a number but a ratio of two lengths related to a circle: π = *Circumference* : *Diameter*. The next activity examines some historically interesting ratios whose measurement involved keen observations and basic mathematics.

Navigating through Measurement in Grades 9–12

Moon Ratios

Goals

- Simulate a historical determination of an important ratio
- Use similar triangles to estimate the ratios of quantities
- Understand an ancient use of a ratio of time to make an important measurement

Materials and Equipment

For each student—
- A copy of the activity sheet "Using a Distance-to-Diameter Ratio"
- A copy of the activity sheet "Using a Ratio of Time"

For each group of three or four students—
- A tape measure (either a retractable carpenter's tape measure or a traditional 36-inch tape measure)
- A coin
- A protractor (optional)
- Centimeter grid paper (optional; template provided)

For the class—
- A Styrofoam ball (4–6 inches in diameter)

pp. 131–32; 133

You can print grid paper for this activity from the template "Centimeter Grid Paper" on the CD-ROM.

Discussion

From the surface of the earth, the disks of the sun and the moon appear to be the same size. This is so because their *visual angles* are roughly equal. Figure 4.4 shows how to measure a visual angle. What angle does the sun or the moon appear to subtend to a person standing on the earth's surface? Though ancient sources sometimes credit Thales with first measuring the visual angle of the sun or moon as 1/2°, Archimedes attributes this very good estimate to Aristarchus (ca. 230 B.C.).

The activity "Moon Ratios" has two parts. In the part 1, "Using a Distance-to-Diameter Ratio," students estimate the ratio of the diameter of an object to its distance from them. They relate this ratio to the size of the visual angle. In part 2, "Using a Ratio of Time," students explore ancient methods of comparing intervals of time elapsed during two parts of a total lunar eclipse and use the ratio to compare the diameters of the earth and the moon.

The *visual angle* of an object is the angle that the object displaces in someone's field of vision. This angle is formed by two rays of light traveling from opposite points of an object to the center of the eye.

If the visual angle of the moon were smaller than that of the sun, we would never witness a total solar eclipse.

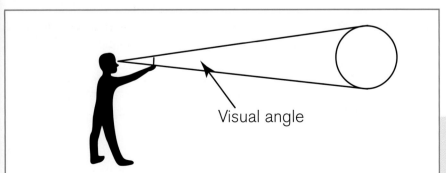

Visual angle

Fig. **4.4.**

Measuring the visual angle of a distant object

Part 1—"Using a Distance-to-Diameter Ratio"

To set the stage for part 1 of Moon Ratios, place a Styrofoam ball in the classroom where everyone can see it from a distance of at least 10 feet. Assign your students to groups, and give each group a coin and a tape measure.

As the activity sheet explains, ancient astronomers invented various methods to determine the ratio of the earth-moon distance to the diameter of the moon. They recognized the value that the ratio could have in helping them make related measurements. Students use the coin, tape measure, and Styrofoam ball to simulate a method that early astronomers used for determining this significant ratio. Like the early astronomers, who then used the ratio to help them estimate the moon's visual angle, the students use the ratio that they obtain from their simulation to help them estimate the visual angle of the Styrofoam ball in their classroom.

Explain *visual angle* to your students, making a drawing on the board like the one in figure 4.4, if such a picture would be helpful. Part 1 of Moon Ratios challenges the students to use the coin and the tape measure to determine the ratio of the ball's distance to its diameter, as well as the visual angle of the ball.

The activity challenges students to make as few direct measurements with the tape measure as possible. Necessary measurements include the diameter of the coin and its distance from the student's eye.

Students can determine the ratio in question by holding the coin so that it exactly "blocks out" the ball from their field of vision, an arrangement that they can represent graphically as in figure 4.5. If they make direct measurements of the diameter of the coin (shown as *AB*) and its distance from their eyes (*d*), then they can readily compute the ratio

$$\frac{AB}{d}.$$

Since $\triangle ABE$ is similar to $\triangle DCE$,

$$\frac{AB}{d} = \frac{DC}{k},$$

or the ratio of the ball's diameter (*DC*) to its distance from them (*k*).

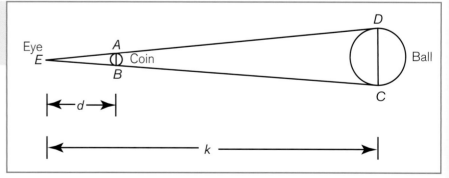

Fig. **4.5.**

Using similar triangles to find ratios of distances

Students can find the visual angle ($\angle AEB$) of the coin and the ball by using their measurements of *AB* (the diameter of the coin) and *d* (the coin's distance from their eye) to draw the triangle *ABE* on grid paper. Then they can use a protractor to measure $\angle AEB$. However, if they have studied trigonometry, they will not need to measure. They can compute the visual angle as

$$2 \times \arctan\left(\frac{0.5 \times AB}{d}\right).$$

Parts 1 and 2 of Moon Ratios both ask students to focus on the relative error of their measurements. The solutions illustrate how students might go about computing the error.

Part 2—"Using a Ratio of Time"

In part 2 of Moon Ratios, students build on their work in part 1 while learning about an important use that ancient astronomers made of their observations from lunar eclipses. Astronomers timed full eclipses of the moon and used their data to compare the diameters of the moon and the earth. Figure 4.6 sketches a simplified version of the situation. The shadow of the earth is actually shaped like a cone, with a taper of one-half of a degree—roughly the visual angle of the sun. Working with the time that the moon was in the earth's shadow, and considering the taper of the shadow-cone of the earth, ancient astronomers reckoned the earth's diameter to be approximately three-and-one-half times the diameter of the moon. Hirshfeld (2004) gives a brief and highly readable account of the way in which Aristarchus considered these issues.

Early astronomers measured two intervals of time:

1. The time from when the moon began entering the earth's shadow to when it was entirely within the shadow (the moment of full eclipse)
2. The time from full eclipse to when the moon completed its emergence from the earth's shadow

Students are asked to consider how someone could use these two measurements of time to compare the diameters of the moon and the earth, assuming that the sun's rays are parallel as they strike the earth. After your students have completed their analysis of problem 1, you should point out that ancient astronomers adjusted their estimate to reflect their assumption that the earth's shadow tapered by one-half of a degree—the visual angle of the sun from the earth.

If your students have not previously encountered the term *relative error,* the discussion in "About This Book" (pp. vii–xii) can help you explain it to them. We usually estimate the accuracy of a measurement by the ratio *Maximum possible error in the measurement* : *Measurement*.

See "The Triangles of Aristarchus" (Hirshfeld 2004; available on the CD-ROM) for an account of how Aristarchus approached the issues involved in measuring the relative sizes of the sun, moon, and earth.

Fig. 4.6.

A diagram showing the moon inside the earth's shadow during a lunar eclipse, with the sun's rays assumed to be parallel

These measurements enabled early astronomers to estimate the diameter of the earth as approximately 3.5 times the diameter of the moon. In the activity, students use the polar diameter of the earth (8400 miles; a measurement not so different from that obtained by Thales), to calculate the moon's diameter as approximately 8400/3.5, or 2400 miles.

Thus, the moon's radius is about 1200 miles. A diagram such as that in figure 4.7 illustrates that the distance from the earth to the moon is equal to

$$\frac{\text{radius of moon}}{\tan\left(\frac{1}{4}°\right)} \approx 275{,}000 \text{ miles.}$$

Fig. 4.7.

A diagram showing the distance from an observer (O) on Earth to the center of the moon (the radius of the moon is shown as r_m)

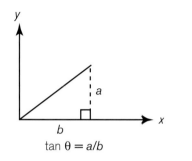
$\tan \theta = a/b$

In the Cartesian plane, the slope of a segment from the origin is the tangent of the angle θ formed by the segment and the positive *x*-axis (see the illustration in the margin). High school students know that each angle θ has a unique slope. If students have a table of slopes such as table 4.1, they can complete this part of Moon Ratios without trigonometry.

Table 4.1.
Slopes of Selected Angles of Inclination (0° ≤ θ ≤ 90°)

Angle (θ°)	Slope	Angle (θ°)	Slope	Angle (θ°)	Slope
0	0	1	≈ 0.0175	80	≈ 5.6713
1/15	≈ 0.0012	5	≈ 0.0875	82	≈ 7.1154
1/6	≈ 0.0029	10	≈ 0.1763	84	≈ 9.5144
1/5	≈ 0.0035	30	≈ 0.5774	86	≈ 14.3007
1/4	≈ 0.0044	45	= 1	88	≈ 28.6363
1/3	≈ 0.0058	60	≈ 1.7321	89	≈ 57.2899
1/2	≈ 0.0087	75	≈ 3.7321	90	Undefined

Assessment

When measurement activities call on students to design and carry out their own measurement methods, assessment should focus on several key elements that are similar to attributes that Kerr and Lester (1976) describe. Try to answer the following questions about your students' work:

- What assumptions did the students make about the objects that they were measuring and the kinds of measurements that they hoped to obtain?
- What measurement strategies or designs, including measuring instrument(s) and mathematical model(s), did the students come up with to accomplish the measurements?
- Did the students' strategies lead to appropriate solutions?
- How well did the students carry out their strategies, including making reliable and accurate measurements and calculations?

- How well did the students assess the potential errors in their measuring process and their final conclusion?

Where to Go Next in Instruction

It is one thing to worry about the smallest unit of measurement allowed by the instrument that one person is using and the rounding off that he or she necessarily does in making the measurement. However, a different concern is that there is often considerable variation in the measurements obtained when many people use the same instrument to measure the same phenomenon individually.

In the next activity, students independently measure the same angle. They discover that the measurement process becomes more complicated when deciding on a value for the angle and determining the precision of the value bring statistical concepts and methods into play. Like the activity Approximately Speaking in chapter 1, the next activity, How Far Is the Sun, acquaints students with the intersection of measurement with data analysis.

See "An Error Analysis Model for Measurement" (Kerr and Lester 1976) for additional ideas on helping students analyze measurement error.

Navigating through Data Analysis in Grades 9–12 (Burrill et al. 2003) helps teachers introduce high school students to statistical concepts and methods.

See "The Archaeological Dig Site: Using Geometry to Reconstruct the Past" (Moyer and Hsia 2001) and "Designing the Dynamic Domino Race" (Bremigan 2002) for additional activities that engage students in issues related to replicated measurements, precision, and accuracy. Both articles appear on the CD-ROM.

How Far Is the Sun?

Goals

- Simulate a historical measurement of an important ratio
- Investigate variability in multiple measurements of the same event
- Recognize that small errors in measurements can yield large relative errors

Materials and Equipment

For each student—
- A copy of the activity sheet "Figuring Out the Phases"
- A copy of the activity sheet "Angling for the Distance"

For the class—
- A bright light (a flashlight or halogen lamp)
- A large protractor
- Transparent tape

For each group of three or four students—
- A Styrofoam ball (4–6 inches in diameter)
- A wooden dowel (approximately 2 feet long and 1/8–1/4 inches in diameter)
- A tape measure (either a retractable carpenter's tape measure or a conventional 36-inch tape measure)
- A plastic drinking straw
- Pencil and paper or geometric drawing software (optional for students who have not studied trigonometry)

If supplies are limited, groups can share these materials, or the whole class can work together.

pp. 134–36; 137–39

You can print a large protractor from the template "Protractor" on the CD-ROM. You can stiffen the paper protractor by taping or gluing it to cardstock if you wish.

Discussion

The activity How Far Is the Sun? consists of two parts—"Figuring Out the Phases" and "Angling for the Distance." Students can work in groups on both parts. However, if your students have not studied the phases of the moon before, you might decide to introduce the topic by having the whole class work together on the first part of the activity.

Part 1—"Figuring Out the Phases"

In part 1, students use a light source as the sun and a Styrofoam ball as the moon in a simulation of the phases of the moon. They measure the simulated sun-moon-earth angles (∠SME) in different phases. When they simulate the half-moon phases, they discover the 90-degree sun-moon-earth angles that proved useful to the Greek mathematician and astronomer Aristarchus (ca. 310–230 B.C.) in estimating the distance from the earth to the sun.

Have each group of students push a dowel firmly into a Styrofoam ball. Mount a bright light at least five feet from the floor at one end of the classroom and dim the other lights. Several students can perform the experiment at the same time if they stand far enough apart. Each student should stand eight to ten feet from the light source, holding the dowel with an arm fully extended and raised so that the Styrofoam ball is just a little bit higher than the student's head. (See fig. 4.8.) Tell the students to face the light and look steadily at the ball.

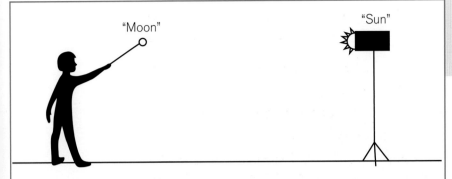

Fig. **4.8.**

The setup for the simulation of the phases of the moon in the activity How Far Is the Sun?

The students should turn slowly in place, continuing to look at the ball and rotating 360 degrees in a counterclockwise direction, imitating the earth's rotation. As they turn and watch the light shift on their Styrofoam ball, they will witness a good display of the various phases of the moon. Each student in the class should have a chance to perform the experiment.

Understanding the changing phases of the moon will pose a challenge to students whose only acquaintance with the effects of the moon's motion coupled with the earth's motion has been a casual inspection of pictures of the moon's phases in a science book. Figure 4.9 shows the situation from two points of view: (a) a vantage point in space from which all the phases look the same (in the example, they all look like half-moons) and (b) a vantage point on Earth, from which all the phases look different.

Share this illustration with your students if you think it will help them grasp the phases of the moon, the motions and spatial relationships that cause them, and the terms that we conventionally use to describe them. The activity sheet explains *gibbous, waxing,* and *waning* to facilitate students' work with the simulation and with the chart on "Figuring Out the Phases." The activity asks students to estimate the sun-moon-earth angle at various phases. By considering the symmetries of the situation, students can reason that a sun-moon-earth angle of 90° occurs exactly when there is a half moon.

However, the problem is that a moon appears to be a "half-moon" over an interval of angles, not just when the sun-moon-earth angle is 90°. Students are asked to estimate intervals of angles that they might associate with a full moon, a gibbous moon, a half-moon, a crescent moon, and a new moon.

Part 2—"Angling for the Distance"

In the second part of the activity, students use a simulation of the earth-moon-sun configuration in which the moon appears to be a half-moon from Earth and *m∠SME* is 90°. This configuration gave Aristarchus a way to make his estimate of the distance from the earth to the sun. The activity lets students explore Aristarchus's method while showing them the potential of small errors in direct measurements to produce large errors in derived values.

Chapter 4: Classic Examples of Measurement

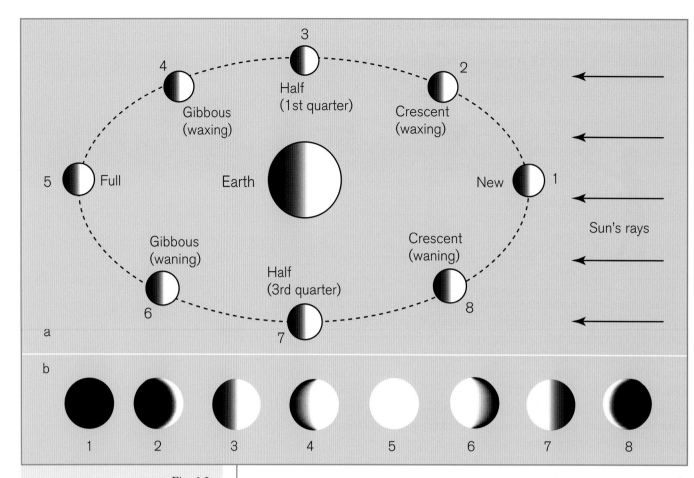

Fig. **4.9.**

Phases of the moon as seen (a) from space and (b) from the earth

You will need to guide your students in setting up one desk or table for the simulation, as shown in figure 4.10:

(1) Mount a Styrofoam ball (the "moon") at a corner of a rectangular table or desk.

(2) Place a light (the "sun") about three feet above the floor pointing at the ball.

(3) Position the light so that it is collinear with one of the table edges that touch the ball.

(4) Measure to set the light so that it is at least 10 feet away from the ball.

(5) Using tape, mount a protractor on the other corner of the table that is adjacent to the "moon." (This corner serves as the "earth.")

(6) Position the protractor so that its center is on the corner (the "earth") and its 0° mark is on an edge of the table and points toward the ball (the "moon").

Fig. **4.10.**

The setup for a simulation for measuring the sun-earth-moon angle; S stands for "sun," M, for "moon," and E, for "earth"

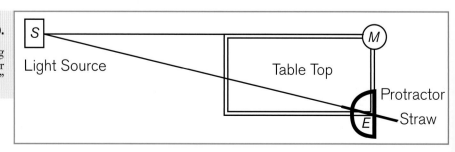

Call groups of students forward one by one, and have the students take turns placing the straw on the corner of the table with the protractor and looking through it to find the center of the light source. Suggest to your students that they use a hand to anchor the straw at the center of the protractor as they work. Once they have taken care to position their straw correctly, they should record the angle that it makes with the table's edge containing the 0° mark of the protractor. This is $m\angle SEM$. Students should not share their results with one another until everyone in all the groups has recorded his or her measurement.

There will be variation in the angle measurements that the students record. Direct your students to work either as a class or in groups to pool their data and determine an angle measurement that represents the central tendency of the data set. The activity sheet also calls on the students to give an estimate of the uncertainty interval of the measurement. The solutions suggest several methods that students might use to make this estimate, including doubling the standard deviation in the measurements. This technique gives a statistically sound estimate of the uncertainty interval.

In question 3(*b*), students with a background in basic trigonometry can reason that the cosine of $\angle SEM$ equals the ratio of lengths *EM/ES*. They can use this fact and their measurement of *EM* to calculate an estimate of the length *ES*.

Alternatively, students who have not yet studied trigonometry can use the angle that they measured to make a right triangle that is similar to that in their simulation. They can construct the triangle either manually, with paper, pencil, and protractor, or electronically, with geometric drawing software. Reasoning from the similarity of the triangles, they can conclude that the ratio of the adjacent side of the measured angle in their triangle to the triangle's hypotenuse is the same as *EM* : *ES*. Equipped with this insight, they will be ready for you to tell them that the ratio has a name, the cosine of the angle, and that their calculators can easily produce this ratio for any angle.

Two sources of error that students usually mention in their error analyses are the measurement of the sun-earth-moon angle and the determination of the precise moment when the moon is a half-moon. Interestingly enough, these were the principal sources of difficulty for Aristarchus, as well. The last part of the activity explores the implications of these sources of error. Students should realize that although Aristarchus's method was mathematically sound and in many respects brilliant, it was also of little practical use because of the huge difference that a small error in measurement of an angle will make in the cosine ratio when the angle is close to 90°.

Aristarchus used 87° although the actual angle is close to 89°52′. Thus, Aristarchus concluded that the distance to the sun was about 19 times the distance to the moon: $1/\cos(87°) \approx 1/.05234 \approx 19$. The true ratio is closer to 400 to 1.

Mathematics historian Otto Neugebauer argues that Aristarchus knew that his angle measurements were very inaccurate and probably not valid even as observational data. Neugebauer conjectures that Aristarchus wrote his treatise *On the Sizes and Distances of the Sun and Moon* purely as a mathematical exercise, to demonstrate "the power of a

---

Statistics texts often assume that the errors in repeated measurements of the same phenomenon are normally distributed around zero. To explore this aspect of measurement, see Navigating through Data Analysis in Grades 9–12 *(Burrill et al. 2003).*

Students compute an estimate for the earth-sun distance that depends on Aristarchus' measure of ∠*SEM*, or 87°. They compare this estimate to the estimate that NASA customarily uses: 93 million miles. For a discussion of this modern estimate, see http://nco.jpl.nasa.gov/glossary/qu.html.

> "All students should ... analyze precision, accuracy, and approximate error in measurement situations."
> (NCTM 2000, p. 320)

mathematical approach to astronomical problems" (Neugebauer 1975, p. 643).

The activity closes with a question about the relative error of the measurement, or, as it is sometimes called, the measurement's accuracy. Even round-off error that results from the resolution of the measurement instrument can lead to a large relative error in the context of this activity. For example, if ∠SEM is measured to be 89°, with a maximum round-off error of .5°, then the relative error in the cosine measurement alone, comparing cos(89°) to cos(89.5°), is

$$\frac{\cos(89°) - \cos(89.5°)}{\cos(89°)} = \frac{.0175 - .0087}{.0175},$$

or 50 percent.

Assessment

Perhaps the most important skill to check in this activity is your students' ability to calculate estimates of the relative error in their measurements. The angle measurements that students make are not very reliable even though they are made with a protractor. The problem is not so much the round-off error that is due to the resolution of the instrument but rather the variation that is intrinsic to the method of observation. Even the different ways in which students place their straws must be considered in estimating the potential error or uncertainty in the measurements. This variation is similar to the unreliability that not knowing exactly when a half-moon occurs introduces into the measurements of the sun-earth-moon angle.

Your students should understand that when they are estimating the uncertainty in a measurement, they must often consider other factors besides round-off error. They should realize that in such situations multiple observations and some basic understanding of statistical methods can be helpful in coming up with an appropriate value and estimate of error.

Conclusion

Chapter 4 has illustrated ways of using the intertwined histories of mathematics and science as valuable sources for measurement and mathematical modeling activities. Chapter 5, the book's final chapter, illustrates the wealth of possibilities that another area—modern technology—offers for activities that can enrich students' understanding of measurement and the role of mathematical modeling.

NAVIGATIONS SERIES

GRADES 9–12

NAVIGATING through MEASUREMENT

Chapter 5
Measuring with Advanced Technology

According to the NCTM Technology Principle, "Electronic technologies—calculators and computers—are essential tools for teaching, learning, and doing mathematics.... When technological tools are available, students can focus on decision making, reflection, reasoning, and problem solving." (NCTM 2000, p. 24)

We measure times, distances, weights, numbers of books, volumes, velocities, calories, and other quantities with a purpose in mind. Many times that purpose is to understand some physical, psychological, sociological, or economic phenomenon. In the world of industry, a box may serve a specific commercial purpose for which its size is important. Yet, manufacturers may also want to know how a change in one or more of the box's dimensions would affect the weight of the box or its volume. To achieve such a purpose fully and effectively, the manufacturers may need to make many measurements and study the behavior of one measured quantity in relation to another. Investigations of this type often call for mathematical models that can relate the measured quantities and account for—predict—how the quantities will change in relation to one another.

The pervasive, rapid innovation in the technology of measurement has made such investigations of measurements not only possible and cost-effective in the worlds of industry and science but also feasible and appropriate in the classroom. Understanding the strategies and the technologies for such investigations should not be knowledge that students acquire only after they leave the school and enter the workplace. Teaching our students how to apply technology to enrich measurement is an attainable goal in the mathematics classroom. In fact, according to *Principles and Standards for School Mathematics* (NCTM 2000), learning how to use technology for measurement should be an important aspect of the mathematical preparation of all students.

Scientists have developed numerous electronic measuring devices, many of which are both inexpensive and classroom-appropriate. In addition, the Internet offers a profusion of resources for all kinds of

> "Scientists rely on technology to enhance the gathering and manipulation of data.... The accuracy and precision of the data, and therefore the quality of the exploration, depend on the technology used." (National Research Council 1996, p. 176)

> For a basic introduction to data analysis and curve fitting on a graphing calculator, see "Using Graphing Calculators to Model Real-World Data" (Holliday and Duff 2004; available on the CD-ROM).

measurements. The challenge for teachers is to understand these technological advances and to determine how to help students use them, along with the measurements that they produce, to study the behavior of natural phenomena intelligently. This chapter presents four activities in which students investigate real-world settings with powerful mathematical tools and measurements that modern technology has made available.

In the first activity, Starbucks Expansion, students collect data from the Web to investigate the increase in the number of Starbucks coffee shops over time. The second activity, Golf Ball Boogie, guides the students in developing a mathematical model of the motion of a golf ball as captured on a strobe-light photograph. The third activity, Bouncing Ball, follows up on Golf Ball Boogie by helping students use an electronic measuring device to gather real-time data on the bouncing of an actual ball. In the fourth and final activity, Most Like It Hot, students use an electronic probe to investigate changes in the temperature of a cup of hot coffee.

In presenting these and similar activities to our students, we serve two masters—science and mathematics. Research in the physical, biological, social, and decision sciences requires proficiencies with mathematical tools and techniques. Work in our own discipline calls for a deep and broad understanding of mathematical concepts.

To use such modeling activities successfully, we should always focus on two questions:

- "What is the mathematics in the activity?"
- "How important is the use of this mathematics?"

Students should be able to complete all the activities if they are familiar with some basics of data analysis and have some acquaintance with curve fitting on a graphing calculator. In the last two activities, students use some increasingly sophisticated mathematical ideas that ultimately lead to the study of differential equations in AP Calculus.

Starbucks Expansion

Goals

- Collect measurement data from the Internet and investigate its reliability
- Use measurement data to create mathematical models
- Represent data and models graphically and use the results to make predictions

Materials and Equipment

For each student—
- A copy of the activity sheet "Starbucks Expansion"
- Paper, pencil, and a graphing calculator (or a computer and spreadsheet software)

For each group of three or four students—
- Access to the Internet

pp. 140–41

Discussion

We can obtain and make good use of all sorts of data from the Internet if we only know where to look for the information and how to evaluate its quality. Is the source a governmental agency, an educational entity, a nonprofit organization, a commercial enterprise, or a private individual? These varied sources have different purposes and standards of accuracy and objectivity. Does the source have a vested interest in the accuracy or possible interpretations of its information? Has someone gathered the data in a rigorous and objective fashion, with care for completeness and consistency?

These may be unusual considerations for students and teachers in a mathematics class, but answering such questions is very important. As teachers, we must learn to be skeptical about data that come from unfamiliar sources and avoid applying mathematics to such data without regard to their reliability and accuracy. It is essential that we teach our students to be equally cautious and critical.

In this activity, students investigate the expansion of the Starbucks Coffee Company after searching the Internet for year-by-year counts of the number of Starbucks stores in operation. They can retrieve the data electronically, but recording them manually, with pencil and paper, is also fine. What measurements your students use and what models are appropriate will depend on what you want your students to get out of the modeling process.

Students look at the growth rate of the number of stores as a measure derived from the annual counts. Using the number of stores and the rate of growth, they develop a mathematical model of the company's expansion. Is the expansion linear, or is it exponential? Students make a scatterplot of their data and see if they can find and graph a line of best fit (see fig. 5.1). They then consider models of exponential growth and search for an exponential function that fits the data (see fig. 5.2).

Students can find a suitable set of Starbucks data on the Web at http://www.starbucks.com/aboutus/timeline.asp. They can also use Google.com or another search engine to find other data about the number of Starbucks stores.

Fig. **5.1.**
A linear regression model for Starbucks data

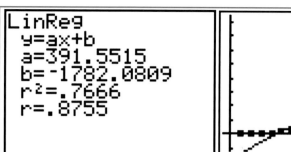
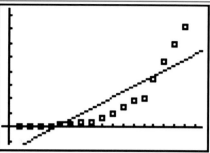

Fig. **5.2.**
An exponential regression model for Starbucks data

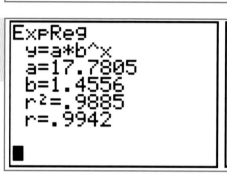
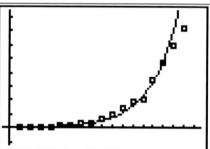

The logistic curve
$$y = \frac{100}{(1 + 5e^{-2x})}$$
has the following graph:

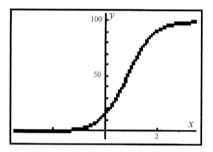

Note that *y* approaches 100 as *x* becomes large, and *y* approaches 0 as *x* approaches negative infinity. Logistic models are useful for situations in which a population begins growing in an exponential fashion but then experiences a decrease in its growth rate as the population reaches the carrying capacity of its environment.

See "Teaching the Logistic Function in High School" (Stephens 2002; available on the CD-ROM) for ideas on how to teach eleventh and twelfth graders about this important mathematical model.

As with all measurements, the students must consider the accuracy of the numbers and the reasonableness of the model that they are basing on these numbers. They must also evaluate the usefulness of their model in predicting the future expansion of the Starbucks Coffee Company.

You can extend this activity by having your students consider other measures of a company's growth, such as the rate at which the company's net sales or net profit is increasing. Students can again retrieve appropriate data from the Internet. You can help precalculus students analyze the relationship between the exponential model $y = a \cdot b^x$ and the logistic model

$$y = \frac{c}{1 + a \cdot e^{-bx}}$$

to shed light on the company's growth patterns. In a calculus course, you can discuss the rate of change and the differences between the derivatives of the mathematical models, and your students can usefully investigate the growth rate in the data.

Assessment

In assessing your students' work, consider the following four questions:

1. Did the students increase their awareness of data that are available on the Internet and recognize the need to be critical in examining the sources and accuracy of the information?

2. Did the students take into account how they would be using their data when they compared different models for the same data?

3. Did the students understand that a model must give reasonable results in the range of values for which someone might use it?

4. Did the students assess the appropriateness of their model in relation to the characteristics of the situation and the data, regardless of the exactness of the fit?

With adjustments, you can use these questions in assessing your students' work on the subsequent activities in this chapter as well.

In addition to measuring your students' understanding in these areas, you should check their skills in using the technology appropriately. Also be sure to consider their grasp of the technical details of the mathematical models and the graphs generated by the calculators.

Where to Go Next in Instruction

Unlike Starbucks Expansion, in which the students use data retrieved from the Web, the next activity requires students to generate their own measurement data from a photograph. Because of the unusual nature of the photograph, students must carefully decide what measurements to take, what models to develop, and how to interpret the results of their models.

Golf Ball Boogie

Goals

- Generate measurement data from a photograph
- Investigate measurement issues that photographs, films, and videotapes present
- Develop a model of motion from a still photograph

Materials and Equipment

For each student—
- A copy of the activity sheet "Golf Ball Boogie"
- A copy of the photograph "Golf Ball Bounce" by Harold E. Edgerton
- A centimeter ruler
- A graphing calculator or access to spreadsheet software

For each group of three or four students—
- A golf ball
- A tape measure (calibrated in both inches and centimeters, if possible)

Discussion

In this activity, students investigate the motion of a bouncing golf ball by using a high-tech photograph in combination with low-tech measurements that they make themselves with a ruler. Noted photographer Harold E. Edgerton used a stroboscope to produce the multiple-exposure photograph "Golf Ball Bounce" (see figure 5.3), which the activity explores.

Students use a standard centimeter ruler to measure the distance of each golf-ball image from the surface on which it is bouncing. They then come up with a model of the motion of the ball.

Students must make many strategic measurement decisions to decipher

pp. 142–44

A larger reproduction of Harold Edgerton's remarkable photograph "Golf Ball Bounce" is available on the CD-ROM that accompanies this book.

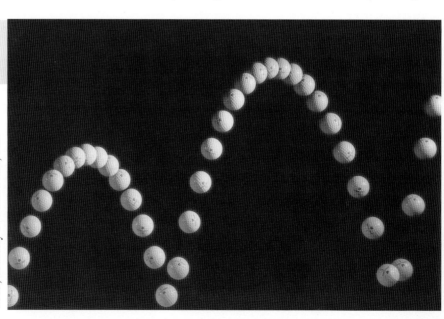

Fig. **5.3.**

"Golf Ball Bounce" (1940) by Harold E. Edgerton

critical information from this photograph. Their choices can lead to wide variation in their measured data and, hence, in the mathematical models that they produce.

Because Edgerton's photograph shows a scaled representation of the situation rather than a life-sized image, the measurements that the students make from the photograph are not the actual measurements that they need for the activity, which calls for "realistic" estimates. To get realistic data, students must measure the size of an actual golf ball and determine a scale for the photograph.

Students must also determine the direction in which the golf ball in the photograph is traveling and the location of the surface on which it is bouncing. In addition, they need to decide what direction counts as "vertical" in the photograph and whether to measure from the center, the top, or the bottom of the ball.

All the advanced technology is in the photograph; the students use only simple technology to obtain measurement data to analyze the phenomenon that the photograph shows. This low-tech approach to a high-tech image can lend itself to many similar investigations. Your students can conduct a variety of measurement explorations with other photographs or with images from films or videotapes.

You can extend Golf Ball Boogie by asking your students to make other measurements. You might ask them to find one or more of the following:

- The vertical velocity of the ball at the end of the middle bounce
- The type of function that models the horizontal motion of the ball
- An estimate of the gravitational constant (32 ft/sec^2 or 9.8 m/s^2) from their data for the middle bounce

If your students do this last exercise, make sure that they specify the units.

Although the activity sheet gives the interval of the strobe light's flashes, you can challenge your students to use the gravitational constant to compute the time between flashes. Again, the units of the students' measurements will be important.

Assessment

Perhaps the most challenging task that students face in this activity is scaling. Take time and care in assessing your students' skills in this important area. Scaling demands proportional reasoning and skill in moving back and forth between measurements in the model and measurements in the photograph.

Students need to understand that errors in scaling and in units can invalidate their models. For example, students who measure the circumference of the actual golf ball in inches will compute an incorrect scale factor for its relationship to the golf-ball image unless they convert the diameter of their golf ball from inches to centimeters before computing the scale factor.

Also assess your students' proficiency with the process of modeling with a quadratic polynomial. Focus particularly on their understanding of the way in which units can sometimes help them interpret the coefficients in the model. For example, in a quadratic model that gives the vertical distance of the ball from the surface in terms of time, $d(t) = a_2 t^2 + a_1 t + a_0$,

Golf balls comply with some standards but have no specific regulation size (see www.thedesignshop.com/history.htm).

Many sites on the Web feature examples of Harold E. Edgerton's photographic studies of objects in motion (see www.susqu.edu/art_gallery/seeing/seeing.htm). Some of these photographs might inspire your students to conduct other investigations like that in Golf Ball Boogie.

"All students should ... use unit analysis to check measurement computations."

(NCTM 2000, p. 320)

the terms $a_2 t^2$, $a_1 t$, and a_0 all represent vertical distances in centimeters, the units that the activity requests. Since t represents time in seconds, the units of a_2 must be centimeters/second^2, indicating that a_2 is a measure of acceleration. Likewise, the units of a_1—centimeters/second—indicate that a_1 is a measure of velocity, and the units of a_0—centimeters—show that a_0 is a distance.

It is also very important for students to understand that the modeling process relies on key assumptions. What is the size of the golf ball? Where is the surface on which it is bouncing? Does the surface slope, or is it level? Does the photograph actually show where the ball hits the surface? To assess your students' understanding of the role of assumptions in their work, you might ask them to write a journal entry in which they list three essential assumptions. Have them explain the impact of these assumptions on their measurements and models. Tell them to evaluate the consequences of a change in any of their assumptions.

Where to Go Next in Instruction

In Golf Ball Boogie, students gather their own data the old-fashioned way—with a ruler. Working from a photograph forces them to make assumptions, which they must analyze as they defend the accuracy, reliability, and validity of their measurements, methods, and models. The measurement process in the next activity is much more direct, but it is nonetheless challenging, as students attempt to measure and model the phenomenon of an actual bouncing ball.

Bouncing Ball

Goals

- Use a motion detector to collect data
- Investigate the collection rate and effective range of the motion detector
- Create and analyze an algebraic model for the motion of a bouncing ball

Material and Equipment

For each student—

- A copy of the activity sheet "Bouncing Ball"
- A graphing calculator

For each group of three or four students—

- A fully inflated rubber ball (approximately 6–10 inches in diameter)
- A motion detector (for example, a TI Ranger or a Vernier motion detector)
- A cable to connect the motion detector to a graphing calculator

If supplies are limited, groups of students can share these materials, or the whole class can work together to collect the data.

pp. 145–47

Discussion

In this activity, students investigate the vertical motion of a large ball as it bounces freely after a drop. Figure 5.4 shows the equipment for the experiment. The students hold the ball at a height of about four feet while also holding a motion detector above the ball at a height of about six feet. They drop the ball and activate the motion detector at the same time. The motion detector, which the students can connect to a graphing calculator by a cable, measures the height of the ball above the floor at equally spaced instants in time over an interval of four seconds. The data are displayed on the calculator.

Once the students have succeeded in collecting usable distance and time data and graphing their data, they select individual parabolic shapes from different bounces to analyze, and they use their calculators to derive a quadratic polynomial to fit their separate sets of data. The students compare their results and look for common features in their quadratic functions.

Note that the students are obtaining measurements for a large ball (6 to 10 inches in diameter), so air resistance is a factor. Also note that though the motion detector in the activity is a TI Ranger, students can use other sorts of motion detectors, including a Vernier motion detector.

After the students have become familiar with the operation of a sonic motion detector like the TI Ranger, ask them to determine the accuracy of its distance and time measurements. They can do this in a variety of ways. For example, they can use the motion detector to measure the distance of an object from a surface—preferably a flat surface—

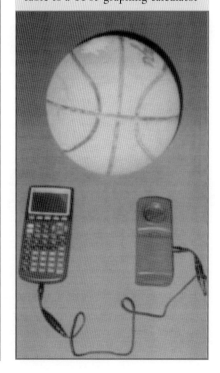

Fig. **5.4.**

Equipment for the activity Bouncing Ball, including a TI Ranger motion detector connected by a cable to a TI 83 graphing calculator

"Calculator- and computer-based measurement instruments facilitate the collection, storage, and analysis of real-time measurement data."
(NCTM 2000, p. 321)

For the activity Bouncing Ball, a sample data set, whose graph is shown on the activity sheet, appears on the CD-ROM.

How does the Global Positioning System (GPS) work? Mission Mathematics II: Grades 9–12 *(House and Day 2005) investigates this question in detail in "Unit 4: Global Positioning System" (pp. 114–154).* Navigating through Geometry in Grades 9–12 *(Day et al. 2001, pp. 31–34) also discusses the topic.*

whose distance from the device (as determined by a measuring tape, perhaps) is known. Students should also determine the range of distances over which the motion detector can accurately measure distance. The user's manual specifies the range of the TI Ranger as from 1 to 6 meters. Even though your students are not conducting these checks rigorously to scientific standards, it is important that they understand the issues of accuracy, reliability, and validity and recognize that these considerations are also relevant to applications of high-precision electronic devices.

This activity might remind you of the classic high school textbook problem that asks students to compute the total distance that a ball travels, assuming that it bounces straight up and down, starting from a known height, h_0, and rebounding to a height equal to two-thirds of the previous rebound height. The solution is a simple application of geometric sequences and series; the height of the nth rebound is

$$h_n = \left(\frac{2}{3}\right)^n h_0.$$

The activity Bouncing Ball asks students to prove that this equation does in fact give the rebound height that the ball reaches between bounces, assuming that it always rebounds to a height equal to two-thirds of its previous height. A rigorous proof would involve mathematical induction. However, you can help your students construct an informal proof by giving them a hint. Suggest that they list the first few elements of the sequence and look for a pattern that continues.

Students can work from

$$h_n = \left(\frac{2}{3}\right)^n h_0$$

and calculate the distance that the ball travels by using formulas for summing geometric series. It will often turn out that a geometric sequence is a good model for the rebound heights in your students' data sets. The activity asks the students to find a geometric sequence that fits their measured rebound heights.

The students must consider the following question: "If the rebound heights of your ball form a geometric sequence, should your ball still be bouncing?" This question can produce interesting responses from your students. On the one hand, those compelled by what they see when looking at the ball may say, "No, it stopped." On the other hand, those compelled by the assumption that the geometric sequence truly models the heights of the rebounds may say, "But the ball always rebounds and so should never stop." The solutions explain how to resolve this conflict by observing that though the model implies that the ball will bounce infinitely many times, the time between bounces is rapidly decreasing, and thus the total time associated with the bounces is finite.

The proliferation of classroom-accessible electronic measuring devices gives rise to important extensions. Chapter 1 showed how significant mathematics—the real number line—could be embedded in the conception and design of even the simplest measuring device—the ruler. The same is true of all modern electronic measuring devices. How does the Global Positioning System (GPS) determine location? How does a sonic motion detector turn echoes into distances? How does it achieve the accuracy that it claims? (Ask your students to use the

Internet or consult an engineer to find out how such sonic motion detectors work.)

The Ranger's Ball Bounce program performs some data conversion. The motion detector records the distance of the ball from the detector, yet the graph appears to record the distance of the ball from the floor. How does the program make this conversion? As an extension, you might ask students to recover the original data from the converted data set and to write the conversion function.

The main point here is that we frequently overlook the mathematics that humans embed in their measuring tools. This "hidden" mathematics can be a very rich source of meaningful classroom activities with clear real-world applications.

Assessment

It is important to assess whether your students are using the sonic measuring device correctly and appropriately. For example, do they understand that the distances that they attempt to measure need to be within the range of the detector? Do they understand the scaling in the graphical representations that the detector produces of the data?

As in Golf Ball Boogie, you should also assess your students' understanding of the units of the coefficients in the quadratic model. Correctly identifying the units is essential though often challenging to students.

A common misconception of students looking at a distance-time graph of the first few bounces of a ball is that the time interval between bounces is constant. This misunderstanding can lead these students to argue that if the bounce heights form a geometric sequence, then the ball can never stop bouncing. If the students will only listen to the sounds of the successive bounces, they will understand that the time between bounces is decreasing.

In fact, as the sample data in the solutions show, if the heights of the bounces form a geometric sequence, then so do the times between the bounces. For many balls, a geometric sequence is not a bad model for these times, and you can tell your students they are listening to a geometric sequence!

Where to Go Next in Instruction

The last two activities have directed students to the quadratic polynomial as an appropriate mathematical model of the motion of a ball. Students have focused on details of the measuring device, units, scaling, and other issues. The following activity, Most Like It Hot, places more emphasis on analyzing data to determine an appropriate mathematical model.

See "Forever May Only Be a Few Seconds" (O'Connor 1999; available on the CD-ROM) for a good discussion of the mathematics of the problem of the bouncing ball.

Most Like It Hot

Goals

- Use electronic devices to collect measurements
- Analyze the measurements by graphing and building models

Materials and Equipment

For each student—
- A copy of the activity sheet "Most Like It Hot"
- A graphing calculator

For each group of three or four students—
- An electronic temperature probe (or a digital thermometer)
- A data collection device (Texas Instruments Calculator-Based Laboratory [TI CBL 2] or Vernier LabPro, for example)
- A cable to connect the data collection device to a graphing calculator
- A Styrofoam cup

For the class—
- A source of hot water (above 150° Fahrenheit, or 66° Celsius)
- Instant coffee

If supplies are limited, groups of students can share these materials, or the whole class can work together to collect the data.

Discussion

Most people like their coffee hot, and they want it to stay hot for quite a while after it is poured. One way to ensure that coffee will remain hot for a time is to pour it when it is scalding. However, as one fast-food corporation had the misfortune to discover, this practice can lead to lawsuits if the coffee spills and causes burns. Nevertheless, a specialty coffee business like Starbucks might want to know how quickly coffee cools so that it could design cups and set pouring temperatures that would deliver a product that would maximize the customers' satisfaction.

This activity gives your students an opportunity to explore the phenomenon of cooling coffee—and a few other things, as well. The students collect data on the temperature of a cup of hot coffee over a short interval of time. Figure 5.5 shows the equipment for collecting and analyzing the data. The activity is set up for a thermocouple-based temperature probe and a graphing calculator, but it can work just as well with a digital thermometer and a watch that measures seconds. Either way, the students use sophisticated technology to gather their data. They then create and describe graphical models of the cooling of the coffee. The solutions provide a sample data set.

The students should determine the units for time and temperature and use those units in labeling their graphs. Although the calculator automatically scales the graph, the students must address the issue of scaling when they determine the distance between the tick marks. They

pp. 148–50

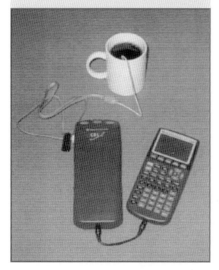

Fig. **5.5.**

Equipment for the activity Most Like It Hot, including a temperature probe, a TI CBL 2 data collection device, and a TI 83 graphing calculator can approximate the nonlinear data points that they have graphed with functions that they define themselves, or they can use their calculators to find curves that seem to be the best fit for the temperature data. A function from the exponential family is likely to provide a better fit than one from the linear family.

A discussion with your students about the difference between curve fitting and model building may be appropriate. Students need to keep in mind that they are measuring for the purpose of understanding a particular phenomenon—the cooling of coffee in a Styrofoam cup in their classroom. Strictly from the point of view of curve fitting, they can always find a polynomial of suitably high degree that will exactly fit the data generated by the temperature probe and the other electronic measuring devices. However, such a polynomial will almost always be a very poor model of the phenomenon under study, since it will be overly sensitive to the errors that are inherent in the measurements. Consequently, a polynomial obtained by curve fitting will often "overfit" the data.

Nevertheless, students need to understand that in developing a model they can also take the ideal of simplicity too far. For example, by using a straight line to model a nonlinear phenomenon like cooling, students risk being insensitive to important trends in the measurements, such as the fact that the hotter the coffee is in relation to the temperature in the room, the faster the coffee's temperature drops. Figure 5.6 shows scatterplots of the sample data with linear and exponential models.

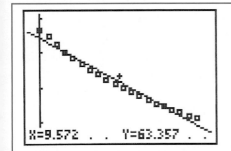 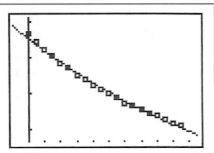

Fig. **5.6.**

Graphing calculator screens showing scatterplots of the sample data for Most Like It Hot with linear and exponential models

Once students have their measurements, they can analyze and model them in different ways, depending on whether or not they have studied calculus. Precalculus students can use difference equations to model the behavior of the temperature of the cooling coffee with respect to time. They can use the standard modeling concept *Future = Present + Change*, or *Next = Now + Change*, where the change is determined by physical principles. In particular, you might suggest that they consider Newton's Law of Cooling, according to which the change in the coffee's temperature will be directly proportional to the difference between the ambient temperature of the room and the temperature of the coffee, as measured by the probe. Letting T_n represent the nth temperature reading of a cup of coffee and assuming that the time interval between readings is constant, we get the difference equation $T_n - T_{n-1} = k[T_{ambient} - T_{n-1}]$, where $T_{ambient}$ is the temperature of the room and $T_n - T_{n-1}$ is the change in temperature from the $(n-1)$th reading to the nth reading.

By contrast, calculus students will readily see that Newton's Law of Cooling leads to the differential equation

$$\frac{dT}{dt} = k \cdot (T_{ambient} - T),$$

with T standing for temperature, and t standing for time. Both equations yield an exponential decay model for the temperature data. We will demonstrate this for the difference equation.

Let $U_n = T_n - T_{ambient}$. Substituting into the difference equation $T_n - T_{n-1} = k[T_{ambient} - T_{n-1}]$ and simplifying, we get $U_n = U_{n-1} \cdot (1-k)$. Thus, $U_1 = U_0(1-k)$, $U_2 = U_1(1-k) = [U_0(1-k)](1-k) = U_0(1-k)^2,\ldots$.

The pattern in the first terms of this sequence of equations yields $U_n = U_0 \cdot (1-k)^n$. Using $U_n = T_n - T_{ambient}$ again, but substituting now for U_n, we get $T_n - T_{ambient} = [T_0 - T_{ambient}](1-k)^n$, or $T_n = T_{ambient} + [T_0 - T_{ambient}](1-k)^n$. Since T_n is a decreasing sequence, then $0 < k < 1$. Consequently, the model predicts that the temperature will decline exponentially.

Assessment

Three of the questions suggested for assessment of students' learning from the activity Starbucks Expansion are appropriate for assessing learning from the current activity:

- Did the students take into account how they would be using their data when they compared different models for the same data?
- Did the students understand that a model must give reasonable results in the range of values for which it might be used?
- Did the students assess the appropriateness of their model in relation to the characteristics of the situation and the data, regardless of the exactness of the fit?

In addition, consider the following questions in evaluating your students' work on Most Like It Hot:

- Do your students understand the coefficients of their models?
- What sense do your students make of the functions and coefficients that they obtain when they use the regression options on their calculators?
- When students are not using the calculators' regression options, how do they determine the coefficients in the models that they try?
- Do your students know what units to associate with the coefficients in their models?

After your students have completed the activity, you may want to assign them a new, though similar, task, which you can use as a performance assessment of their understanding. Having your students gather data and build a model to explain the warming of a glass of ice water can provide you with such an opportunity. You can have your students gather the data as a class. Then, with everyone working with the same data, each student can construct an appropriate model and compare it with his or her preferred model from the cooling experiment.

Conclusion

The activities in this chapter have shown that the phenomenal growth of classroom-accessible technology calls on us to consider the increasingly complex role of measurement in the school curriculum. The technology invites us to broaden and extend our treatment of measurement in the classroom, incorporating sophisticated devices and showing both what they can do and the mathematics on which they rest.

The final section of the book, "Looking Back and Looking Ahead," takes a brief look at the ground that students and teachers should have covered together in grades 9–12, as well as a quick look forward to some of the work with measurement that students may do as they move beyond high school. The book closes by suggesting some of the implications for teachers and students of mathematics.

NAVIGATIONS SERIES

GRADES 9–12

NAVIGATING through MEASUREMENT

Looking Back and Looking Ahead

The activities in this book illustrate the richness and variety of measurement as a strand in the mathematics curriculum in grades 9–12. These explorations highlight one of the book's main ideas, that throughout history, the need for more sophisticated measurements has continually led to the discovery of more sophisticated mathematics. In the same way, hands-on problem solving and investigating in measurement can lead students to a deeper and more complex understanding of mathematics.

However, high school students' experiences of measurement in their science and mathematics classrooms can be confusing unless science and mathematics teachers collaborate on their goals and expectations. Traditionally, science and mathematics teachers have approached measurement from distinct points of view, which have influenced their perceptions of many aspects of the topic, ranging from their interpretations of the nuances of the terminology to their ideas about what is important.

As we have seen, mathematics and science teachers often apply different methods for rounding off and reporting the uncertainty or error inherent in measurements of continuous quantities. Some of these discrepancies result from the slightly different meanings that mathematics and science have traditionally given to the terms *accuracy* and *precision*. Mathematics and science teachers must not ignore these and other differences that are rooted in the histories of their disciplines.

Yet, even if teachers address these differences directly and openly in their classrooms, students cannot appreciate them unless they have first grappled with the phenomenon of measurement itself, in many different settings, from the earliest grades to grade 12. Moreover, activities such as those in this book demonstrate to high school students that they have not yet mastered measurement in its full complexity.

Students in grades 9–12 begin to realize, for example, that coping with the uncertainty in measurement can lead to the application of increasingly subtle statistical tools. They appreciate, perhaps for the first time, the usefulness of the methods of advanced data analysis. They discover the necessity for such methods in many scientific investigations—especially those that place a premium on replicated measurements and reproducible results. This realization prepares students for the more rigorous approaches that scientific investigations use at the college level. Simultaneously, students may discover a personal interest in pursuing statistical studies beyond high school.

Likewise, activities such as those in the book help students understand that ideas from advanced branches of mathematics, like trigonometry and calculus, have concrete, immediate, and highly practical uses in measurement. Students discover firsthand that advanced mathematical ideas "come in handy" in a great variety of real-world measurement contexts. Students also begin to recognize that they can think of measurement as a function, whose properties they can formalize and study in an exact fashion. This discovery prepares students to encounter the formal definitions of mathematical measures, including distance, area, and volume, in mathematical studies in college and beyond.

We have written this book with the goal of serving important needs of the mathematics and science disciplines. This aim rests squarely on our highest hope for the book—that it will illustrate the potential of measurement to connect students to the world around them. We will have achieved this purpose if the ideas and activities that we have presented help you guide your high school students in discovering and taking away with them a fundamental truth: the mathematics that they learn in the classroom equips them with invaluable tools for understanding the rich and complex world in which they live. For no less a purpose, *Principles and Standards for School Mathematics* (NCTM 2000) identifies measurement as one of the major strands in the mathematics education of all children, from prekindergarten through grade 12.

Navigating through Measurement

Grades 9–12

Appendix
Blackline Masters and Solutions

Counting on Commensurability

Name _____

Quest for the Golden Ruler–Part 1

We know that a segment of length $\frac{5}{6}$ feet can be measured as 10 inches, and a segment of length $\frac{3}{4}$ meters can be measured as 75 centimeters. The Pythagoreans, a group of early mathematicians (ca. 550 B.C.), supposed that the counting numbers would always suffice for measurement if the ideal units could be found, as in the above examples.

1. Suppose that you have a blank ruler and a line segment that is $\frac{271}{360}$ as long as your ruler. Also suppose that to measure the segment, you divide your blank ruler into 360 equal units that you will call "jarboos."

    a. How long is your line segment, as measured in jarboos? _____

    b. Suppose that you have a second line segment that is $\frac{5}{6}$ as long as the ruler. How long is this second segment if you measure it in jarboos? _____

2. Suppose that you have two different line segments, one $\frac{5}{7}$ as long as your ruler from step 1, and the other $\frac{4}{11}$ as long as the ruler.

    a. Why would it not be convenient for you to measure these two segments in jarboos?

    b. What is the smallest number of equal units into which you would need to divide the ruler if you wanted to measure both of these segments with whole numbers of units? _____

3. Suppose that you have two other line segments, one of which is $\frac{5893}{3798}$ as long as the ruler, and the other is $\frac{1379}{482}$ as long as the ruler.

    a. Could you divide the ruler into yet another set of units that would allow you to measure both of these segments with whole numbers of those units? _____

    b. How would you determine how many units your ruler would have?

You can measure the pairs of line segments in steps 2 and 3 in whole numbers of units as long as you choose the right units. This result means that the line segments in each pair are *commensurable*. The Pythagoreans believed that any two segments are commensurable.

Is this true? Part 2, "That's Irrational," continues the investigation.

That's Irrational

Name _____

Quest for the Golden Ruler—Part 2

Consider a right triangle with two congruent legs that are exactly equal to the length of a blank ruler. Obviously, we could divide the ruler into any whole number of units we wanted, and the legs would be that number of units in length. However, could we divide the ruler into a whole number of units so that the *hypotenuse* would also be a whole number of those units?

In other words, is the hypotenuse of an isosceles right triangle commensurable with the leg? The Pythagoreans thought it must be, if they could only find the right units.

1. Suppose that you divide a ruler into n units called "enths" and make an isosceles right triangle with legs that are exactly equal to your ruler.

    a. How long is a leg of your triangle in enths? _____

    b. What is the length of the hypotenuse as measured in enths? (Use the Pythagorean theorem.) _____

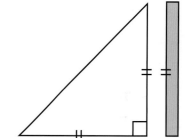

2. Suppose that you can also measure the hypotenuse of your triangle in a whole number of enths—say, m enths. Combine this supposition with your results in step 1 to show that $\sqrt{2} = \dfrac{m}{n}$.

Here the Pythagoreans' conjecture runs into trouble. If the leg and the hypotenuse of this simple triangle really are commensurable, then there must be whole numbers m and n such that the ratio $\dfrac{m}{n}$ can be squared to equal 2. The Pythagoreans did not know of any such ratio—and for good reason. None exists, as you will discover in the remaining steps of this exploration.

3. Assume that there really is a rational number $\dfrac{m}{n}$ in lowest terms that satisfies the equation $\left(\dfrac{m}{n}\right)^2 = 2$. Show that this equation implies that $m^2 = 2n^2$.

4. You can conclude from step 3 that m must be even and m^2 must be a multiple of 4. Why?

5. If m^2 is a multiple of 4, the equation $m^2 = 2n^2$ implies that n^2 must be a multiple of 2 and n must be even. Why?

That's Irrational (continued)

Name _____

6. However, you assumed that $\frac{m}{n}$ is in lowest terms, and in step 4 you concluded that *m* must be even. So *n* would have to be odd. Why?

7. Your conclusions in steps 5 and 6 are both logical results of your assumption in step 3.

   a. Can both conclusions be true? _____

   b. What, therefore, can you conclude about your assumption in step 3?

Super Bowl Shipment

Name _____

Approximately Speaking—Part 1

Suppose that a trucker receives a last-minute request to carry a large shipment of souvenir dolls to the Super Bowl, but he can add only 8,000 pounds to his load without running a risk of failing a highway inspection. The shipment consists of 100 cases, each with 150 dolls reported to weigh 8.4 ounces apiece, plus packing material said to add a pound to the weight of each case.

1. Assume that the reported weights of the dolls and the packing material are exactly correct.

   a. Compute the total weight of the doll shipment.

   b. Does the total added weight come in under the 8,000-pound limit? _____

   c. Is it very likely that the reported weights are, in fact, exactly correct? _____

2. Assume that whoever reported the weight of a doll as 8.4 ounces was measuring to the nearest tenth of an ounce.

   a. What is the range of the possible weights that a doll fitting this description might have?

   b. Rounding these possible weights to the nearest hundredth of an ounce, what is the largest weight that a doll might have? _____

3. Assume that whoever reported the weight of the packing material as 1 pound per case was measuring to the nearest pound.

   a. What is the range of the possible weights of the packing material?

   b. Rounding these possible weights to the nearest tenth of a pound, what is the most that the packing in a case might weigh? _____

4. Use the weight of a doll from step 2 and the weight of the packing material from step 3.

   a. Recompute the weight of the shipment.

   b. Is the total weight under the 8,000-pound limit? _____

   c. If you were the driver, which estimate would you use, and why?

Navigating through Measurement in Grades 9–12

Paula's Popcorn Box

Name _____

Approximately Speaking–Part 2

Paula is designing a popcorn box as a homework project. She makes and measures her box. Rounding each dimension to the nearest centimeter, she measures her box as 25 cm × 17 cm × 8 cm. Paula thinks that her measurements will give the correct volume to the nearest cubic centimeter. Do you think she is correct? The following steps will help you find out.

1. Assuming that the dimensions that Paula has determined are correct, compute the volume of the box

2. Suppose that Paula's box actually measured 254 mm × 174 mm × 84 mm. If Paula rounded these dimensions, would they give her the dimensions in centimeters that she reported? _____

3. a. Use the measurements in step 2 to compute the volume of the box.

   b. By how many cubic centimeters does this computed volume exceed the volume that you computed from Paula's rounded measurements in step 1? _____

4. Suppose that Paula's box actually measured 245 mm × 165 mm × 75 mm. If Paula rounded these dimensions, would they give her the dimensions in centimeters that she reported? _____

5. a. Use the measurements in step 4 to compute the volume of the box.

   b. By how many cubic centimeters does this computed volume fall short of the volume that you computed from Paula's rounded measurements in step 1? _____

6. Assume that a popped kernel occupies about 6 cubic centimeters of space.

   a. What are the estimation errors in steps 3 and 5, as measured in popped kernels? (*Hint:* How many more or fewer popped kernels might Paula's box hold than you would have estimated from your result in step 1.)

   b. What should Paula report in her homework as the volume of her box?

Rounding Numbers in a Sum

Name _____

Approximately Speaking—Part 3 (Class Discussion)

Working together as a class, consider the following scenarios and questions:

1. An elevator engineer learns that a certain passenger elevator can safely carry up to 5,000 pounds. He posts a sign in the elevator that says "Maximum capacity 21 people."

    a. If each person in a group of 21 people weighs over 240 lbs., could the whole group ride together safely in the elevator?

    b. The engineer does not worry about this happening. Why?

2. A flight attendant on a small passenger plane walks through the passenger cabin just before takeoff. In moving down the aisle, the attendant ignores all passengers except small children and extra-large passengers. The flight attendant counts each small child as −1 and each extra-large passenger as +1. In all, the attendant counts 6 small children and 8 extra-large passengers, for a total of +2. Since 64 passengers are on board, the flight attendant reports to the flight engineer, "We have 64 passengers + 2." The flight engineer multiplies 64 times 175 and adds 2 times 75 to get a total passenger weight of 11,350 pounds.

    a. Assuming that the average adult weighs 175 lbs., account for this estimate of the total passenger weight.

    b. Does this system of estimation seem reasonable to you?

3. Suppose that you are in the grocery store with a large shopping list, and you want to estimate the total cost of your food. You decide that you will round the cost of everything that you put into your basket to the nearest dollar. For example, if a lemon costs 30 cents, you'll round it to $0, and if a cake mix costs $2.57, you'll round it to $3. By rounding to whole dollars, you'll be able to keep the running total in your head.

    a. The central limit theorem predicts that, under certain reasonable assumptions, the more items you put in your basket, the more likely your running total is to be close—even remarkably close—to the exact total. Why do you suppose this is true?

    b. Try this method of estimating the total cost of your purchases the next time you go shopping, and compare your results with those of your classmates.

The Right Rope

Name _____

Early Measuring Devices—Part 1

One of the most famous examples of a clever tool used by ancient engineers was a simple rope divided into twelve equal segments marked by knots at regular intervals.

Of course, the rope could be pulled tight to form a straight ruler with twelve unit lengths, but it could perform a much more useful trick than that!

1. *a.* Ancient engineers used the rope to construct a right triangle. How did they arrange the rope for this feat?

    b. What part of the triangle do you suppose these engineers found most useful? _____ Why?

2. *a.* What modern device, available at hardware stores, would give present-day carpenters the same feature that you identified in step 1(*b*)?

    b. Which device would provide more reliable measurements—the ancient one or the modern one? _____ Why?

Why Ships Measure Speed in Knots

Name _____

Early Measuring Devices—Part 2

Ancient sailors sometimes used a very long knotted rope to measure the speed of ships at sea. They tied knots in the rope at regular intervals and fastened one end of the rope to a buoy, a floating marker that was more or less stationary. As the ship went out to sea, a member of the crew sat in the stern of the ship with the coiled remainder of the rope and a glass container of sand. The sailor fed out the rope, counting the number of knots that went into the water as the ship moved away from the buoy. After a fixed interval of time, the crew member stopped counting and reeled in the rope and buoy.

1. The glass container of sand was important in this process. What do you suppose the ship's crew used it for, and how did it work?

2. How could the crew use the number of knots to determine the ship's speed?

3. Would this be a valid way for a ship's captain to measure the speed of the ship? _____ Why, or why not?

4. a. How reliable would the measurements attained by this method be?

   b. If two different crew members simultaneously measured the speed with different buoys and ropes, would they get the same answer?

   c. What conditions might affect the consistency of measurements made in this manner?

5. Suppose that the knots were 47.25 feet apart, and the interval of time was 28 seconds.
   a. If the crew member counted exactly 2 knots in 28 seconds, what would the speed of the ship be in feet per second?

   b. Convert this speed to miles per hour.

Navigating through Measurement in Grades 9–12

Why Ships Measure Speed in Knots (continued)

Name _____

6. Convert the speed in step 5(*b*) to nautical miles per hour. (Approximately 1.15 regular miles, also known as "statute miles," equal 1 nautical mile.)

7. In light of your answer in step 6, explain why the sailing term for "nautical mile per hour" is *knot.*

More Measurement Methods

Name _____

Early Measuring Devices—Part 3

It would be virtually impossible to catalog all the clever methods that our ancestors used to get reasonably accurate measurements with relatively crude instruments. By exploring the following questions, you can rediscover a few of these methods. The measurement problems that the questions pose are not necessarily those that interested our ancestors.

1. Using only a long rope and a 12-inch ruler, could you find the depth of a deep well? How?

2. Using only a 12-inch ruler and a pencil, could you find the height of a tall building on a sunny day? How?

3. With a thunderstorm approaching, could you use only the claps of thunder and flashes of lightning, together with your wristwatch, to measure the distance between yourself and the storm? How?

4. Using only a 3-by-5-inch index card and a pencil, could you estimate the area of a rectangular soccer field? How?

5. Suppose that you were a paleontologist and you had only a single femur of a dinosaur. Could you give an approximation of the dinosaur's height? How?

6. If you had a pile of pennies and nickels and a standard sheet of paper (8 1/2 × 11 inches), could you find the circumferences of a penny and a nickel in fractions of inches? How?

Mathematical Goat

Name _____

An owner keeps a goat inside a square, fenced region. The square measures 12 feet on each side. When the goat has eaten all the grass inside the fence, the owner takes the goat outside the fence and ties it to the corner post labeled *P* with an 8-foot rope. You will need a compass and a ruler as drawing tools, and a calculator for computation, as you explore the goat's new grazing region.

1. *a.* Use your tools to draw the new grazing region on the grid. Work carefully.

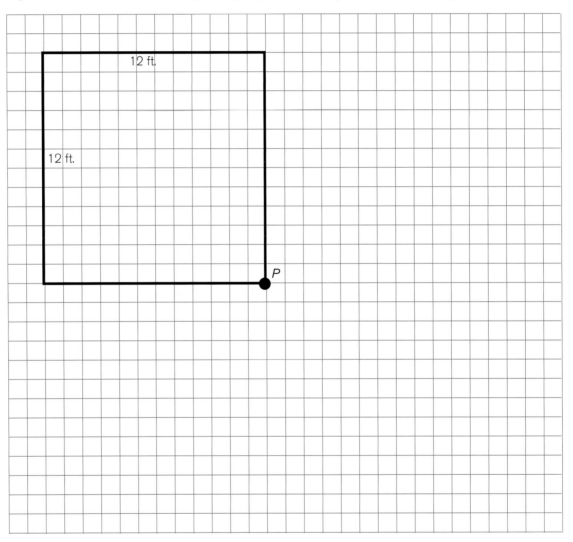

   b. How has the shape changed?

2. *a.* What formula could you use to compute the area of the new grazing region that you have drawn?
   _____

   b. Make the computation, and give your answer both in terms of π and as a decimal.

Mathematical Goat (continued)

Name _____

    c. Compare the area of the new grazing region outside the square to the area of the old square region.

3. a. Count squares as a check for your computation. How many squares (square feet) did you count?
   _____

   b. Compare your count with your computed area. Does your computation seem reasonable?

4. Suppose that the owner replaces the 8-foot rope with a 14-foot rope and again tethers the goat to corner post *P*. Use your drawing tools again, and show the goat's new grazing region on the grid.

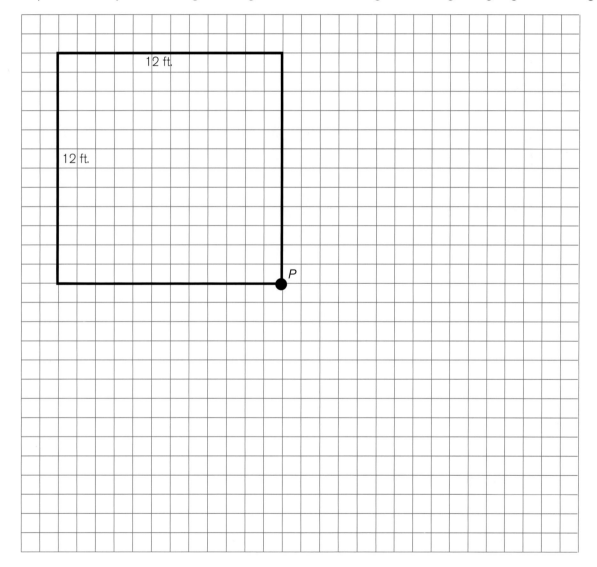

5. a. What formula could you use to compute the area of this new grazing region?
   _____

Navigating through Measurement in Grades 9–12   101

Mathematical Goat (continued)

Name _____

    b. Make the computation, and give your answer both in terms of π and as a decimal.

6. a. Count squares as a check for your computation. How many squares (square feet) did you count?
   _____

   b. Compare your count with your computed area. Does your computation seem reasonable?
   _____

7. Compare the area of the new grazing region created by the 14-foot rope with the region created by the 8-foot rope.

8. Suppose that the owner wants to keep the goat from grazing too much in any single grassy area. The owner decides to continue to use the 14-foot rope but to rotate the tether regularly from one corner post to the next, moving the goat all around the square.

   a. Use your drawing tools again to draw this new grazing region on the grid.

   b. Count squares to estimate the area of this new grazing region. Give your estimate in square feet.
   _____

   c. Can you compute the area of this new region mathematically?

9. Ignore for a moment any practical limits on the length of the rope and the appetite of the goat.

   a. How would the total area of the grazing region change if the owner used a longer and longer rope which he kept moving from corner post to corner post?

   b. If the rope were 1 mile long, what shape could you reasonably say that the grazing area would approximate?

Mathematical Goat (continued)

Name _____

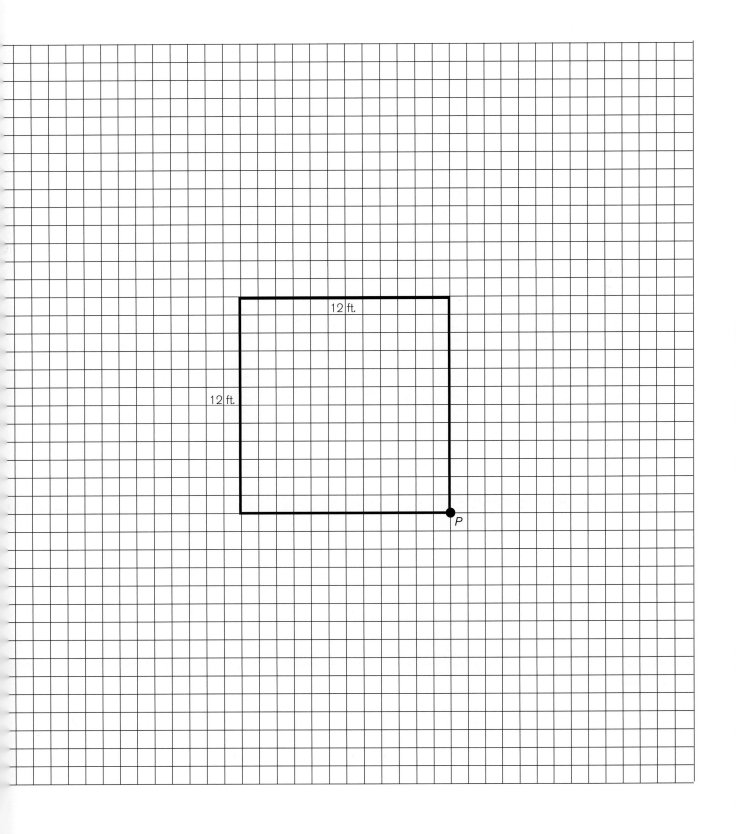

Chip off the Old Block

Name _____

A woodcarver has agreed to cut a sphere and a cylinder from two wooden cubes, each measuring 6 centimeters on an edge, as shown.

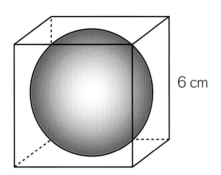
6 cm

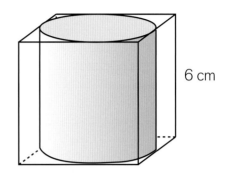
6 cm

1. Write formulas for the volume and surface area of a sphere in term of its radius.

   a. Volume of a sphere _____

   b. Surface area of a sphere _____

2. What is the greatest radius that the woodcarver can achieve in cutting a sphere from one of the 6-centimeter cubes? _____

3. a. Find the maximum volume that the woodcarver can achieve for a sphere cut from one of the 6-centimeter cubes. Express your answer in terms of π and also to the nearest cubic centimeter.

   b. Why is this volume the maximum?

4. Write formulas for the volume and surface area of a cylinder in terms of its radius and height.

   a. Volume of a cylinder _____

   b. Surface area of a cylinder _____

5. Assume that the woodcarver cuts the bases for the cylinder from two faces of one of the 6-centimeter cubes.

   a. What is the greatest radius that the woodcarver can give the cylinder? _____

   b. What is the greatest height that the woodcarver can give the cylinder? _____

Navigating through Measurement in Grades 9–12

Chip off the Old Block (continued)

Name _____

6. a. Find the volume of the cylinder with the dimensions in step 5(a and b). Express your answer in terms of π and also as a decimal to the nearest cubic centimeter.

    b. Do you think that this volume is the maximum that the woodcarver can possibly achieve for a cylinder cut from a 6-centimeter cube? _____ Why, or why not?

7. What is the volume of each of the cubes that the woodcarver has at the beginning? _____

8. How do the volumes that you computed for the sphere in 3(a) and for the cylinder in 6(a) compare with the volume of the cube from which each has been cut? Express the sphere's and the cylinder's volumes as percentages of each cube's volume.

    a. The volume of the sphere to the volume of the cube (percent) _____

    b. The volume of the cylinder to the volume of the cube (percent) _____

9. Use your computations from 3(a) and 6(a) again, this time to determine roughly what fraction of each cube the woodcarver will be throwing away in cutting the sphere and cylinder?

    a. The fraction of the cube discarded for the sphere _____

    b. The fraction of the cube discarded for the cylinder _____

10. Use your computations from 3(a) and 6(a) again, but work this time with the results that you expressed in terms of π. Find the ratio of the volume of the cylinder to the volume of the sphere, and express the ratio in simplest form.

11. a. What is the surface area of the largest sphere that the woodcarver can cut from one of the cubes? Use your formula from 1(b), and express your answer in terms of π and also to the nearest square centimeter.

    b. What is the surface area of the largest cylinder that the woodcarver can cut from the other cube? Use your formula from 4(b), and express your answer in terms of π and also to the nearest square centimeter.

    c. What is the ratio of the surface area of the cylinder to the surface area of the sphere? Express the ratio in simplest form.

Chip off the Old Block (continued)

Name _____

12. *a.* Compare the ratios that you computed in questions 8 and 9.

    b. Would you expect this result any time that you compared the volumes and surface areas of two solids?

    c. Why, or why not? (Archimedes found the same result and was so pleased with his discovery that he asked to have a picture of a sphere inscribed in a cylinder etched on his tombstone.)

13. *a.* Would the results of problems 10 and 11(c) change if the cubes of wood had edges of 2 feet?

    b. State a generalization summarizing your results.

A Base on a Face

Name _____

Cones from Cubes—Part 1

A woodcarver needs to cut a cone from a wooden cube measuring 6 centimeters on an edge. One possibility is shown.

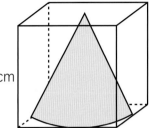

1. Write the formula for the volume of any cone in terms of its radius and its height.

2. If the woodcarver locates the circular base of the cone on one of the faces of the 6-centimeter cube, what are the greatest radius and greatest height that the cone can have?

   a. Radius _____

   b. Height _____

3. Find the volume of a cone with the dimensions that you recorded in step 2.

   a. Express your answer in terms of π. _____

   b. Use an approximation for π, and express your answer from 3(a) as a decimal to the nearest cubic centimeter. _____

4. a. What is the volume of the cube that the woodcarver had at the beginning?

   b. Express the volume that you obtained for the cone in 3(b) as a percentage of the volume of the cube.

5. a. Do you think that this cone has the greatest volume of all possible cones that the woodcarver could cut from a 6-centimeter cube? _____

   b. Why, or why not? *Hint:* Consider inscribing the base of a cone in a regular hexagon formed by a plane slicing through the cube.

Navigating through Measurement in Grades 9–12

Going for the Max

Name _____

Cones from Cubes—Part 2

The woodcarver might be surprised to learn that locating the base of the cone on a face of the cube doesn't give a cone of maximum volume. In fact, the woodcarver could achive the greatest volume by inscribing the cone's circular base in a regular hexagon created by slicing the cube. Let's investigate such a cone.

1. Use the cube pictured here to draw one way in which a plane could slice through a cube to form a regular hexagon. (*Hint:* For a hexagon, the plane must cut all six faces of the cube.)

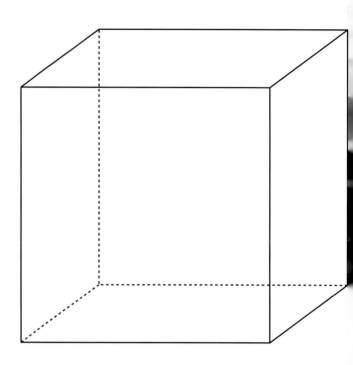

The illustrations below show two ways in which a plane could cut through all six faces of a cube. When this happens, the intersection of the plane and the cube is a hexagon. If the plane also intersects the midpoints of six edges of the cube, the resulting hexagon will be regular.

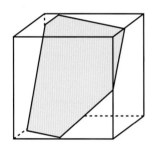 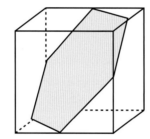

Now suppose that the woodcarver decides to inscribe the base of the cone in one such regular hexagon. Could the woodcarver calculate the dimensions of the new cone without cutting and measuring? Let's investigate.

2. In the cube pictured below, △*PQR* is a right triangle with the right angle at *Q*.

   a. Use the Pythagorean theorem to find the length of diagonal *PR* of the cube.

   b. Give a decimal approximation (to the nearest tenth) of *PR* in centimeters.

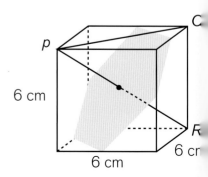

The plane of the hexagon cuts the cube into two congruent halves, bisecting the diagonal *PR* of the cube. Thus, the height of the new cone is half the length of the diagonal of the cube.

108   Navigating through Measurement in Grades 9–12

Going for the Max (continued)

Name _____

3. a. Find the height, *h*, of the woodcarver's new cone.

   b. Give a decimal approximation (to the nearest tenth) of *h* in centimeters.

Since the cone and the hexagon have the same center, the radius of the cone is the perpendicular distance from the center to an edge of the hexagon (see the illustration). In any regular polygon, this distance is called the *apothem*.

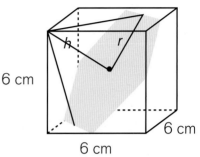

4. Use the Pythagorean theorem or anything else that you know about triangles or hexagons to find the following:

   a. The length of a side of the regular hexagon.

   b. The apothem of the hexagon, which is also the radius, *r*, of the inscribed base of the cone.

5. Use *h* and *r* to find the volume of the woodcarver's new cone.

   a. Express your answer in terms of π. _____

   b. Express your answer to the nearest cubic centimeter. _____

6. a. Express the volume that you have computed for the woodcarver's new cone as a percentage of the volume of the original cube.

   b. Compare this percentage to the percentage that you computed earlier for the cone with its base on a face of the cube.

   c. Do you think that this cone is the biggest one that the woodcarver can cut from a 6-centimeter cube? _____ Why, or why not?

Making a Model

Name _____

Cones from Cubes—Supplement

1. a. Cut out the large square in figure 1.

    b. Cut through to the center of the square on the solid line.

    c. Fold on the dotted lines to form three adjacent faces and the corner of a 6-cm cube.

    d. Tape your construction together.

2. a. Cut out the regular hexagon in figure 2.

    b. Position the hexagon in your model of the cube so that alternate sides of the hexagon rest against the three faces of the cube. All six vertices of the hexagon should be at the midpoints of the edges of the cube.

    c. Tape this new construction together. The inscribed circle on the hexagon will be the base of the cone, which you will make next.

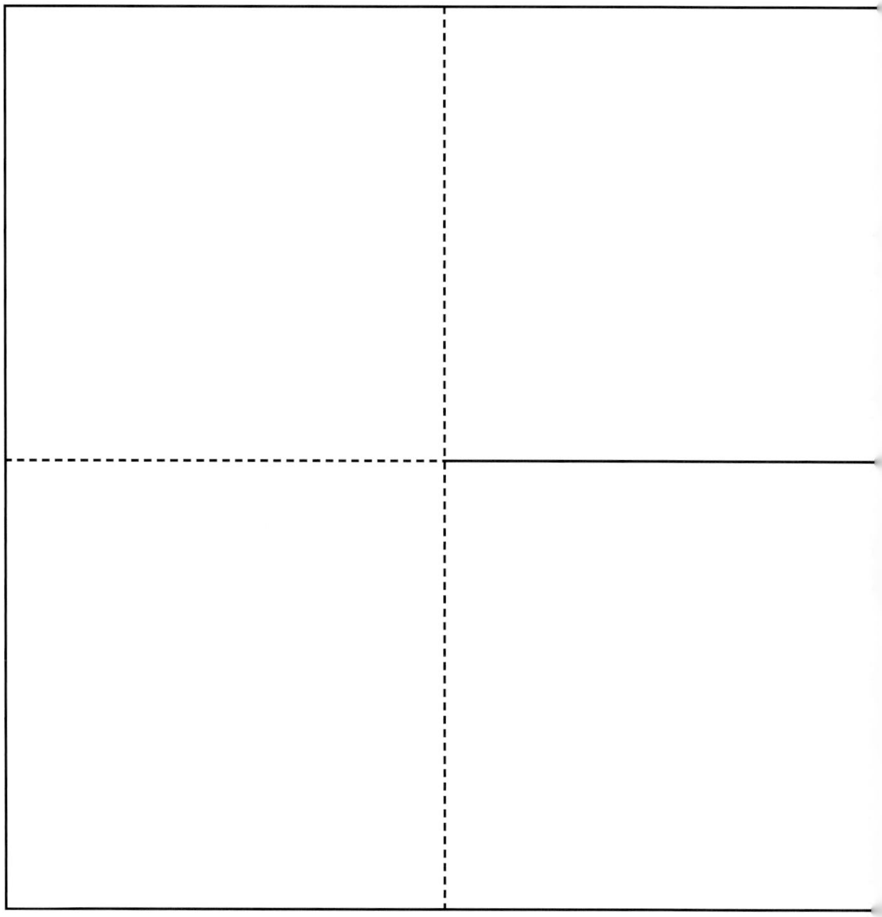

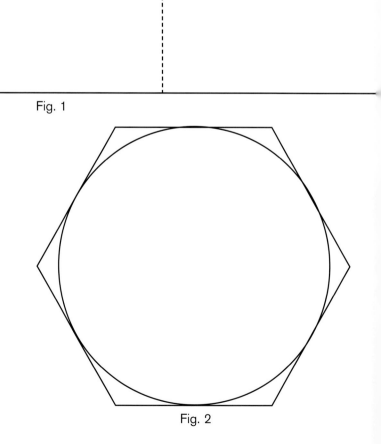

Fig. 1

Fig. 2

Making a Model (continued)

Name _____

3. a. Cut out the circle in figure 3.

   b. Cut through to the center of the circle on the solid line.

   c. Curl the circle to form a cone by overlapping the section marked with the arrow right up to the dotted line.

   d. Tape the cone.

4. a. Verify that the dimensions of the cone are essentially those that you computed in part 2, "Going for the Max."

   b. Position your model of a cone in the cube so that its base matches the circle inscribed in the hexagon. See figure 4.

   c. Check to see that the vertex of the cone is where the eighth vertex of the cube should be.

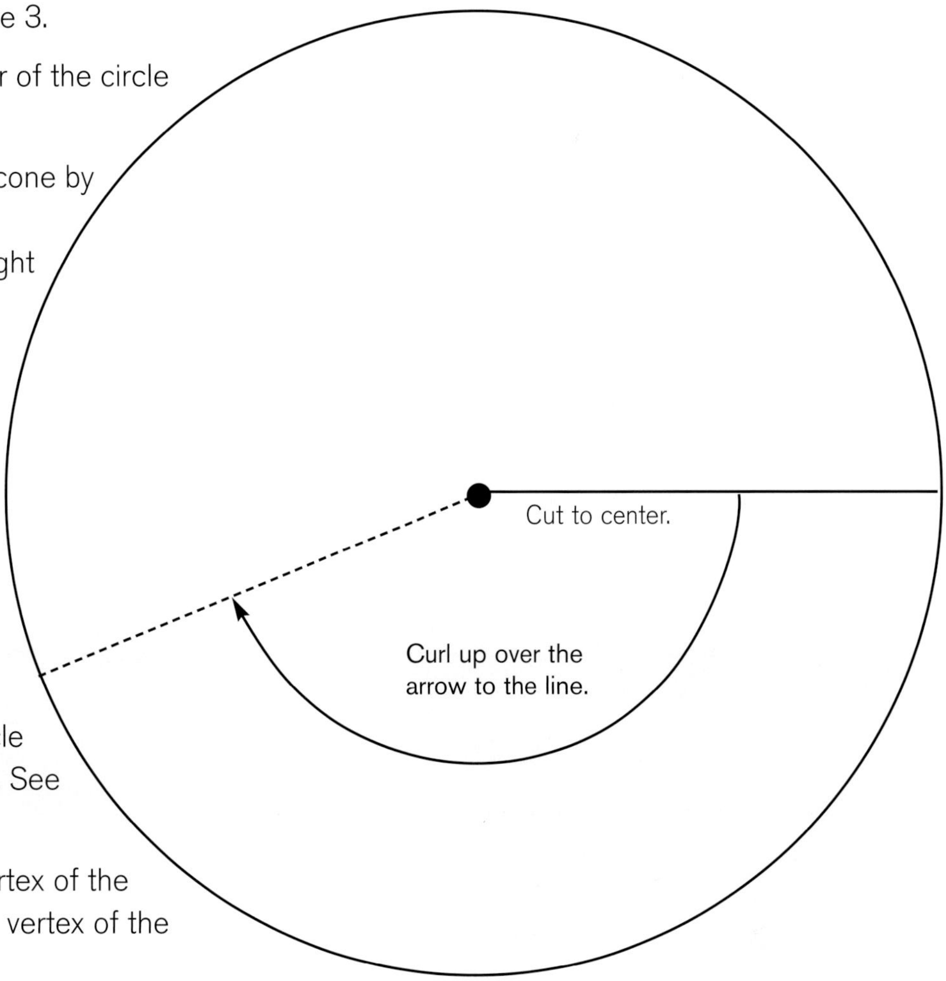

Cut to center.

Curl up over the arrow to the line.

Fig. 3

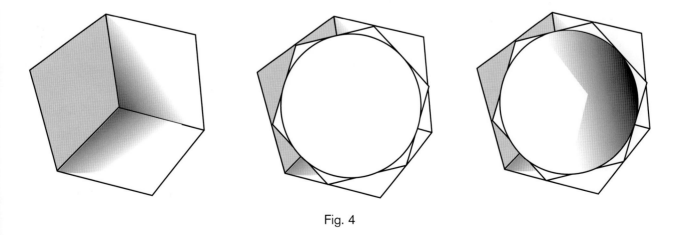

Fig. 4

Iterating on a Plane

Name _____

Measuring a Geometric Iteration—Part 1

Using a ruler, draw a 3-inch square on a piece of paper. Carefully cut out your square, and then complete the following steps:

- Trace an outline of the square on another sheet of paper.
- Measure and cut your original square into thirds each way, creating nine smaller squares.
- Keep five of your smaller squares, and discard the other four.
- Use four of your five squares to fill the corners of your outline, with one square in each corner, and place the fifth smaller square in the center of your outline, with each vertex touching a vertex of one of the four squares in the corners.

The illustration shows the process.

Stage 0

Stage 1

1. *a.* What is the area of the original square at stage 0?

   b. What is the area of the new figure at stage 1?

2. What fractional part of the area of the square at stage 0 is the area of the figure at stage 1?

Iterating on a Plane (continued)

Name _____

3. a. What is the perimeter of the square at stage 0?

   b. What is the perimeter of the new figure at stage 1?

4. When the figure goes from stage 0 to stage 1, does the perimeter increase, decrease, or stay the same? _____

The following rule describes the process that creates the stage 1 figure from a square: "Reduce a square to 1/3 scale, make five copies, and build a new structure by placing those copies in the four corners and the center of a frame formed by the outline of the original square." This rule becomes an *iteration rule* when it includes a step for repeating the process over and over, always using the last figure created to produce the next new one:

Iteration Rule

Reduce a square figure to 1/3 scale and make five copies. Build a new structure by placing those copies in the four corners and the center of a frame formed by the outline of the original square figure. Use the result as the new figure at the next stage, and repeat the process over and over.

An iteration rule has three distinct parts: a *reduction,* a *replication,* and a *rebuilding.* Each time we apply the rule, we create a new and more complex figure.

5. The stage 0 figure consists of one square, and the stage 1 figure has five squares in it. If we apply the rule a second time, how many squares will be in the stage 2 figure? _____

6. a. What is the area of each of the squares at stage 2?

   b. What is the perimeter of each of the squares at stage 2?

The illustrations at the top of the next page show the first four stages of the process. Note that at each successive stage, the figure is more complex than it was at the previous stage. However, your eyes may not see much change, if any, beyond stage 4 or 5. Soon you can only imagine the ever increasing detail as the iteration process repeats over and over.

Iterating on a Plane (continued)

Name _____

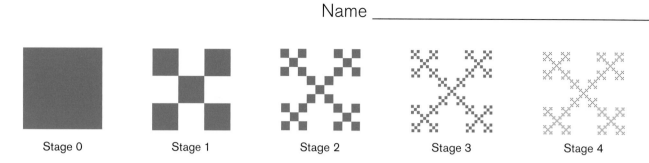

Stage 0   Stage 1   Stage 2   Stage 3   Stage 4

7. Enter the missing values for stages 1, 2, 3, and 4 in the data table. (Leave the column for stage n blank for now.) Give the area and the perimeter for the entire figure at each stage. Use fractions and exponents to express your answers, and look for patterns.

Stage	0	1	2	3	4	n
Number of squares	1					
Total area (in²)	9					
Total perimeter (inches)	12					

From stage to stage, the total number of squares increases as successive powers of 5. At the same time, the total area decreases, converging to 0 in more and more iterations. However, the total perimeter diverges, becoming large without bound. For both the area and the perimeter, there is a constant multiplier that takes us from one stage to the next.

8. *a.* Find the constant multipliers for the rows for the number of squares, the area, and the perimeter in your table.

   b. Try writing the entries for stage n (the last column in the data table) by using increasing exponents on these constant multipliers.

9. Using your calculator, find the stage at which the total number of squares in the figure will first exceed 6.5 billion, the number often given as an estimate of the total number of people in the world.

10. Again using your calculator, find the stage at which the perimeter of the complex figure will first be greater than one mile.

Iterating in 3-D

Name _____

Measuring a Geometric Iteration—Part 2 (Assessment)

Consider a 3-inch cube. Think of its edges as a frame. Suppose that you—

- slice the cube into 27 smaller cubes by cutting its length, width, and height into thirds;
- keep only 9 of the newly formed cubes; and
- place 8 of the smaller cubes in the corners of a frame formed by your original cube, and place the remaining cube in the center of the frame.

The stage 1 result is shown.

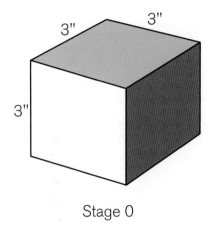

Stage 0   Stage 1

1. a. What is the surface area of the cube at stage 0?

   b. What is the surface area of the new figure at stage 1?

2. a. What is the volume of the cube at stage 0?

   b. What is the total volume of the new figure at stage 1?

The following iteration rule will create this stage 1 figure from a cube:

> **Iteration Rule**
>
> *Reduce a cube to 1/3 scale and make nine copies. Build a new structure by placing eight of these copies in the corners of the frame formed by the outline of the original cube and one in the center of the frame. Use the result as the new figure at the next stage, and repeat the process over and over.*

Navigating through Measurement in Grades 9–12

Iterating in 3-D (continued)

Name _____

Note that the rule has the three distinct parts that an iteration rule must have—a *reduction*, a *replication*, and a *rebuilding*.

3. Use the iteration rule to complete the data table, and analyze any patterns that you find.

Stage	0	1	2	3	4	n
Number of cubes	1					
Total surface area (in^2)	54					
Total volume (in^3)	27					

    a. How does the number of cubes change from stage to stage?

    b. What happens to the surface area as the process continues through more and more iterations?

    c. What happens to the volume as the process continues?

Building Pyramids with Cubes

Name _____

Discovering the Volume of a Pyramid—Part 1

Can you build a pyramid with cubes? Of course you can't build an exact pyramid with cubes, but you can approximate one. The illustration shows two "pyramids" of cubes with square bases.

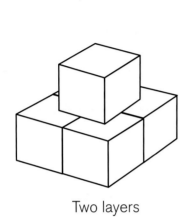

Two layers

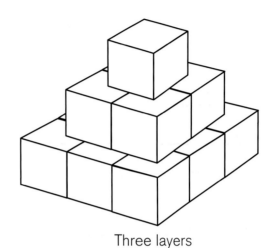

Three layers

Your teacher has given you a set of small cubes. Use the cubes to build a series of such pyramids, each with one more layer of unit cubes than the previous one. Because your pyramids consist of cubes, you can easily calculate each one's volume.

You can also easily compare each pyramid's volume to the volume of a cubic prism with the same base and height. More important, you can use your results to explore the volume of a "real," exact pyramid.

1. What is the volume of the two-layered pyramid of cubes in the illustration? _____

2. What would be the volume of a cubic prism with the same base and height as the two-layered pyramid of cubes? _____

3. Build cube-pyramids, and complete the data table on the next page. For each row, compute the ratio of the volume of the pyramid (the value in the second column) to the volume of the corresponding prism (the value in the third column), and enter the result in the last column.

Building Pyramids with Cubes (continued)

Name _____

Number of layers in the "pyramid" of cubes	Volume of the "pyramid" of cubes	Volume of the corresponding cubic prism	$\dfrac{\text{Volume of "pyramid"}}{\text{Volume of cubic prism}}$
1			
2			
3			
4			
5			
6			
7			
8			
9			
10			
n			

4. *a.* What happens to the ratio in the last column as your cube-pyramids become larger? (Is the ratio increasing, decreasing, getting closer to some number, or staying the same?) _____

_____

b. Why do you think this is happening?

Probing Pyramids with Spreadsheets

Name _____

Discovering the Volume of a Pyramid–Part 2

In "Building Pyramids with Cubes," you briefly explored the volumes of pyramids that you approximated with cubes. The last column in your data table showed the ratio of the volume of each such "pyramid" to the volume of a cubic prism with the same base and height.

As you probably discovered, the pattern in the ratios is hard to detect in a short table. Extending the table by hand quickly becomes tedious. An electronic spreadsheet can easily help you do the work. Your teacher will provide spreadsheet software so that you can pursue the investigation that you began in "Building Pyramids with Cubes."

1. In an electronic spreadsheet, enter the following column headings:
   - Number of layers in the cube-pyramid
   - Volume of the cube-pyramid
   - Volume of the corresponding cubic prism
   - Ratio of the volume of the cube-pyramid to the volume of the prism

   (You will probably want to abbreviate.)

2. Refer to your data table from "Building Pyramids with Cubes." Enter the values from the first two rows into your spreadsheet. (Do not enter numbers for the last column, because you will use a formula to make the spreadsheet calculate these values.)

Number of layers	Volume of "pyramid"	Volume of cubic prism	$\dfrac{\text{Volume of "pyramid"}}{\text{Volume of cubic prism}}$
1	1	1	
2	5	8	

3. Enter a formula for the ratio in the first empty cell in the last column. (Do not simply enter the number 1, the first value for the last column. And remember, formulas in a spreadsheet begin with "=.")

4. Select the formula that you just entered for the ratio, and copy it into the cell below it. Your spreadsheet should now look something like the sample shown.

Number of layers	Volume of "pyramid"	Volume of cubic prism	$\dfrac{\text{Volume of "pyramid"}}{\text{Volume of cubic prism}}$
1	1	1	1
2	5	8	0.625

Navigating through Measurement in Grades 9–12

Probing Pyramids with Spreadsheets (continued)

Name _____

5. The next row of your spreadsheet will contain only formulas. Enter all four appropriate formulas. This step can be challenging. Be sure to use the patterns that you noticed when you began your exploration in "Building Pyramids with Cubes."

6. Select the row of formulas that you just created, and copy the row into the next row. Continue to copy down into more and more rows (you can use any shortcut that your software allows, such as "Fill Down") until your table is very, very long. (It should be long enough that you are sure of a pattern in the last column.)

7. Use the graphing feature of your spreadsheet software to make a graph of the ratio numbers in the last column. (To select the "Number of layers" column and the nonadjacent column that shows the ratio of the volumes, select one column first, and then select the other column while holding down the control key.)

8. What happens to the ratio in the last column of your spreadsheet as the size of the pyramid becomes larger? Will the ratio in this column ever be 0?

9. Use your experience with your "bumpy" pyramids of cubes to make a generalization about conventional, smooth-sided pyramids. (If a pyramid has a base of area B and a height of h, write a formula for its volume.)

Making a Formula

Name _____

Measuring the Size of a Tree—Part 1

Take a look at some trees near your school or your home or in a park. It doesn't matter where the trees are, but they must be real trees! Select a tree to think about and examine as you develop a formula.

1. List attributes that you could measure for your tree—for example, its number of leaves (or needles) or the maximum area of shade that it makes. Try to come up with 20 different attributes.

2. On your list, circle all the attributes that relate to the tree's size.

3. Consider the attributes that you circled for the size of your tree.

    a. Choose at least three attributes whose measures you think would be important to an overall determination of the size of your tree.

    b. List these attributes in the first column in the table.

Attribute	Variable	Unit of measure	How would you make the measurement?

   c. Assign to each attribute a letter of the alphabet that you can use to represent it as a variable in a measurement formula. (For example, logging companies sometimes measure the height of a tree, its girth—the circumference of its trunk—and the spread of its branches to estimate how much timber the tree will yield. They might use the letters h, g, and s as variables to stand for the measures of these attributes.) Enter your variables in the second column in the table.

   d. Decide on a unit of measure for each attribute, and enter it in the third column in the table.

Making a Formula (continued)

Name _____

    e. In the last column of the table, tell how you would measure the attribute. *Note:* Your measurement method must be reasonable, practical, and safe. For example, climbing the tree should not be part of your method of measuring the tree.

4. Devise a formula that uses your variables to calculate the size of your tree. You can make one attribute "count" more or less than another in your formula, depending on how important you think that variable is in relation to the others.

    a. Give your formula.

    b. Explain your formula.

Using Your Formula

Name _____

Measuring the Size of a Tree—Part 2

You have now devised a formula for measuring the size of a tree. Write your formula here:

Do you suppose your formula will really work? That is, can it give you results that you can use reliably and effectively to compare the sizes of trees? Experiment to find out.

1. Find two trees in a setting where you can measure them.

    a. Describe how your trees look and what their location is like. (You might make a sketch of each tree.)

Tree 1	Tree 2

    b. Make your best estimate and circle *bigger* or *smaller* to complete the following sentence:

    Tree 1 is _____ times *bigger / smaller* than tree 2.

2. Perform any measurements that you need for your formula, and record all the numbers and units here. Detail how you performed each measurement. (You might include a picture or a diagram to help you remember what you did.)

    Measurement 1 (for the attribute_____)

Tree 1	Tree 2

Navigating through Measurement in Grades 9–12

Using Your Formula (continued)

Name _____

How did you make the measurement? _____
_____
_____

Measurement 2 (for the attribute _____)

Tree 1	Tree 2

How did you make the measurement? _____
_____
_____

Measurement 3 (for the attribute _____)

Tree 1	Tree 2

How did you make the measurement? _____
_____
_____

Measurement 4 (if needed) (for the attribute _____)

Tree 1	Tree 2

Using Your Formula (continued)

Name _____

How did you make the measurement? _____

_____

_____

3. *a.* Use your formula to calculate the size of each tree. Show all your calculations clearly and carefully.

Tree 1	Tree 2

   b. Complete the sentence again, to reflect your calculations with your formula.

        Tree 1 is _____ times *bigger / smaller* than tree 2.

   c. Compare this newly completed sentence to your sentence in 1(*b*).

Reporting Your Results

Name _____

Measuring the Size of a Tree—Part 3

Reflect on your work with your formula for measuring the size of a tree, and report your results. If possible, use poster board and make a poster that you will present to the class. Otherwise, make a report on clean sheets of paper.

1. Your report should include the following information:
   a. Your formula for the size of a tree and an explanation of how you developed it.
   b. A description of the trees that you measured.
   c. All measurements that you performed on the trees before applying your formula.
   d. Your final calculations of the sizes of the trees, including the units of your measurements.

2. Your report should answer the following questions:
   a. Looking over your results, would you change your formula at all?
   b. What did you learn from this project?
   c. What was your favorite part of this project?
   d. What was the most difficult part of the project?

3. Be sure that you give credit where credit is due. Acknowledge any outside resources that you used, and give thanks to any classmates or other individuals who helped you. If you worked in a group, describe the contributions of all the group members.

4. Check the mathematical content of your report to be sure that it is clear and correct. Include diagrams to illustrate and explain your mathematics.

5. Make sure that your information is clear and well organized. If you are designing a poster, try to make your work colorful and visually appealing.

Assessing a Poster

Name _____

Measuring the Size of a Tree—Part 4 (Optional)

Name(s) of the student(s) whose poster you are assessing _____
_____
_____

Completeness. Check to see that the poster satisfies each of the requirements.

1. The poster gives the following information:

a. A formula for the size of a tree and an explanation of how the student(s) developed it. _____ out of 7 points b. A description of the trees that the student(s) measured. _____ out of 5 points	c. All measurements that the student(s) performed before applying the formula. _____ out of 5 points d. The student's (students') final calculations of the sizes of two trees, including the units of the measurements. _____ out of 7 points

Grade _____ out of 24 points

2. The poster answers the following questions:

a. Looking back, would the student(s) change the formula at all? _____ out of 5 points b. What did the student(s) learn from this project? _____ out of 5 points	c. What was the student's (students') favorite part of this project? _____ out of 5 points d. What did the student(s) think was the most difficult part of the project? _____ out of 5 points

Grade _____ out of 20 points

Assessing a Poster (continued)

Name _____

3. The poster gives credit where credit is due:

(*If the poster was made by a single student*) It acknowledges outside resources and thanks any classmates or other individuals who helped. _____ out of 5 points	(*If the poster was made by a group*) It acknowledges outside resources and describes the contributions of all group members. _____ out of 5 points

Grade _____ out of 5 points

Mathematical Content. Evaluate the mathematical work shown on the poster.

1. The mathematics seems clear.	_____ out of 5 points
2. The mathematics seems interesting.	_____ out of 5 points
3. The mathematics seems correct.	_____ out of 6 points
4. The mathematical explanation uses diagrams.	_____ out of 5 points

Grade _____ out of 21 points

Organization, Layout, and Creativity

1. The information is clear and well organized.	_____ out of 8 points
2. The poster is colorful and visually appealing.	_____ out of 7 points

Grade _____ out of 15 points

Oral Presentation

The presentation was audible, interesting, and easy to follow, and (*if the poster was made by a group*) each group member participated in it.

Grade _____ out of 15 points

Total Grade _____ out of 100 points

If the Earth Is Round, How Big Is It?

Name _____

Eratosthenes was an early Greek mathematician who loved astronomy. In the third century B.C., he devised a method for measuring the circumference of the earth. Eratosthenes measured the angle of the shadow cast by a vertical stick in Alexandria at noon on the summer solstice. This angle (shown as ∠TBA in the diagram; drawing is not to scale) was equal to 7.2°. Eratosthenes didn't measure the angle in degrees, however, since the system of degrees didn't yet exist.

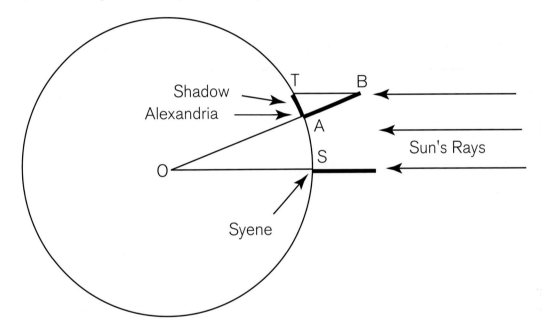

Eratosthenes knew that on the same day, at the same time, due south in the town of Syene (S on the diagram), the sun was directly overhead, so a vertical stick would cast no shadow. He also knew that the distance from Alexandria to Syene was approximately 5000 stades, or about 500 miles. He assumed that the sun was so far from the earth that its rays were parallel. The diagram illustrates the situation.

1. a. If Eratosthenes' measurements are correct, what is the circumference of the earth?

   b. What is the diameter of the earth?

   c. Justify your method for calculating the circumference.

Navigating through Measurement in Grades 9–12

If the Earth Is Round, How Big Is It? (continued)

Name _____

2. *a.* With a Styrofoam ball, a light source (such as a flashlight), and two straight pins, set up a simulation of Eratosthenes' method, and use it to measure the ball's circumference. Record your measurements to the nearest millimeter. *Note:* Because a Styrofoam ball's surface is rough, it may not give you a clean shadow to measure. You can cut a strip from an index card to use as a base for reflecting a shadow.

   b. If a measurement is rounded off to the nearest millimeter, the round-off error is less than 0.5 millimeters. Calculate the possible error in your estimate of the ball's circumference if each of your ruler measurements had an error of 0.5 mm.

3. Bozeman, Montana, is due north of Tucson, Arizona. Suppose that two students, one in Bozeman and one in Tucson, are connected by cell phones. It is noon on a sunny day, and they have gone into their respective school courtyards. By prior agreement, each student has brought a classmate to help hold a meterstick vertically to the ground and measure its shadow while the students on the cell phones communicate to ensure that they are making the measurements simultaneously.

   a. Make a diagram like that in step 1 to illustrate this new situation.

   b. Explain how these two students can use their measurements along with information from a road atlas to make an estimate of the circumference of the earth.

Using a Distance-to-Diameter Ratio

Name _____

Moon Ratios—Part 1

What is the ratio of the earth-moon distance to the diameter of the moon? Why would this ratio be useful to know? Ancient astronomers devised methods for determining it because they realized that it could help them find the distance from the earth to the moon.

To estimate this distance, they needed more than the ratio of the earth-moon distance to the diameter of the moon. They also needed an estimate of the moon's diameter. In part 2, you will explore an ancient method for making this measure. But first, in part 1, you will use a coin and a measuring tape to simulate an early method for determining a distance-to-diameter ratio when you cannot measure the attributes directly.

Your teacher has placed a Styrofoam ball at a distance (at least 10 feet away) in the classroom. How can you use your coin and measuring tape to make an indirect measurement of the ratio of your distance from the ball to the ball's diameter? Don't measure the ball or your distance from it directly! Remember, you are simulating a method for estimating a ratio that early astronomers couldn't ascertain directly—the ratio of the earth-moon distance to the moon's diameter.

1. Use the coin and the measuring tape to determine the ratio of the diameter of the ball to its distance from where you are standing. (Mark this spot with an object or a piece of tape. You'll need this location again in step 5.) Try to make as few direct measurements as possible.

2. Give a mathematical justification for your method, and include a diagram.

3. a. How precise are your direct measurements? Take their precision into account, and determine an error interval for your ratio.

Using a Distance-to-Diameter Ratio (continued)

    b. Give an estimate of the relative error of your ratio.

4. Ancient astronomers used the ratio of the earth-moon distance to the moon's diameter to help them determine the visual angle of the moon—the angle that the moon displaces in the visual field of an observer on Earth. Use your measurements to determine the *visual angle* of the ball. (The visual angle has its vertex at your eye, and its sides are tangent to the ball.)

5. *a.* Now that you have completed all the preceding steps, directly measure the distance from your vantage point to the ball.

    b. Make a direct measurement of the diameter of the ball.

    c. Does the ratio of these measurements fall within the error interval that you calculated in 3(*a*)?

Using a Ratio of Time

Name _____

Moon Ratios—Part 2

Ancient astronomers observed and timed full eclipses of the moon—events in which the moon passes through the center of the earth's shadow. They measured the time that the moon took to become fully eclipsed once it began entering the earth's shadow, and they compared this time to the time from when the moon became fully eclipsed to when it had finished emerging from the earth's shadow (see the diagram).

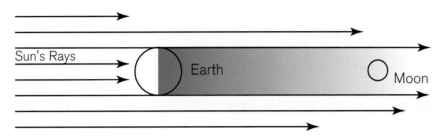

1. Assuming that the sun's rays are parallel when they strike the earth, explain how the ancient astronomers could use their two measurements of time to compare the diameters of the moon and the earth.

2. From their study of eclipses, early astronomers estimated that the diameter of the earth was approximately 3.5 times the diameter of the moon. They also estimated that the visual angle of the moon (the angle that the moon displaces in the visual field of an observer on Earth) was 1/2°. Assume that the polar diameter of the earth is 8400 miles—a measurement that is fairly close to that obtained by the ancient astronomer Eratosthenes. Use these measurements to calculate the distance between the earth and the moon.

3. Radar measurements indicate that the distance between the earth and moon, center to center, is approximately 239,000 miles.

   a. Use this fact to estimate the error in your answer in step 2.

   b. Estimate the relative error in your answer in step 2.

Navigating through Measurement in Grades 9–12

Figuring Out the Phases

Name _____

How Far Is the Sun?—Part 1

How far is the sun from the earth? Aristarchus, an ancient Greek astronomer, used important information gleaned from studying the moon to estimate this distance. In part 1 of the activity, you will simulate the phases of the moon to discover this essential information. In part 2, you will simulate the way in which Aristarchus used the information to estimate the ratio of the earth-moon distance to the earth-sun distance.

A bright light will serve as the sun. For part 1, the light should be mounted at least 5 feet above floor-level and about 10 feet from where you will stand. A Styrofoam ball will serve as the moon. You shaould attach the ball to a dowel by pushing the dowell firmly into the ball. The ceiling lights should be off for the simulation.

1. Holding the free end of the dowel in your hand, fully extend your arm, and raise it so that the "moon" is slightly above the level of your head. Look straight at the "moon," and hold it directly into the "sun-light"—the rays from your light. (See the drawing; notice that your head will simulate the earth.)

2. Turn slowly in place in a counterclockwise direction. Watch the light from the "sun" shift on the "moon" as you turn. Continue turning slowly until you have rotated a full 360 degrees. You are simulating the changes—the phases—that we see in the moon each month (roughly every 29 1/2 days) from our vantage point on earth.

3. Identify the various phases of the moon as you witness them in sequence. Remember that *waxing* means "growing larger" and *waning* means "growing smaller." The term *gibbous* uses the Latin *gibbus* ("humped") and *gibbosus* ("humpbacked") to describe the phases (both waxing and waning) when the moon appears to be larger than half but smaller than full.

Figuring Out the Phases (continued)

Name _____

- New moon

- Crescent moon (waxing)

- Half-moon (1st quarter moon)

- Gibbous moon (waxing)

- Full moon

- Gibbous moon (waning)

- Half-moon (3rd quarter moon)

- Crescent moon (waning)

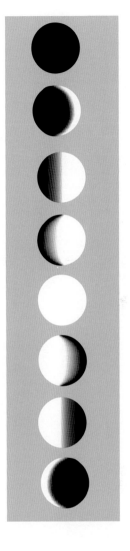

4. *a.* Use a protractor and measure the angles in your simulation between the "sun," "moon," and "earth" ($\angle SME$) for the various phases of the "moon." Complete the chart. *Note:* You may discover a range of angles, rather than a single angle, for each phase.

Moon	New Moon	Crescent Moon (Waxing)	Half-Moon (1st Quarter)	Gibbous Moon (Waxing)
Range of Angles				

Moon	Full Moon	Gibbous Moon (Waning)	Half-Moon (3rd Quarter)	Crescent Moon (Waning)
Range of Angles				

Navigating through Measurement in Grades 9–12

Figuring Out the Phases (continued)

Name _____

    b. Discuss your results with others in your class.

5. Aristarchus believed that the moon appears to be an exact half-moon to an observer on earth when ∠*SME* is a right angle. From your angle measurements, do you think that he was justified in this belief?

Angling for the Distance

Name _____

How Far Is the Sun?–Part 2

Aristarchus believed that the moon appears to be exactly a half-moon when the sun-moon-earth angle ($\angle SME$) is 90°. This right angle was useful to him in estimating the distance from the earth to the sun (*ES*). In this part of the activity, you will simulate the situation in which $m\angle SME = 90°$, to explore Aristarchus's method for estimating the earth-sun distance.

The drawing shows the setup for the simulation:

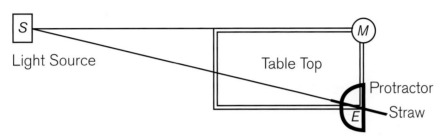

A Styrofoam ball mounted at one corner of a tabletop will represent the moon (*M*). An adjacent corner of the table will represent the earth (*E*). A light source representing the sun (*S*) should be mounted approximately 10 feet away from the tabletop, on a line through *M* and the table's other corner that is adjacent to *M*. A protractor, taped in place with its center at *E* and its 0° mark on the edge of the table, will allow the measurement of angles as though by an observer on the earth.

1. *a.* In your simulation, does the "moon" appear to be exactly a half-moon from the perspective of someone on "earth"? _____

   b. Place a drinking straw across the center of the protractor, as shown in the drawing. Sight through the straw, and find the center of the light source. Measure *EM* and $\angle SEM$ for this configuration, and record your measures below.

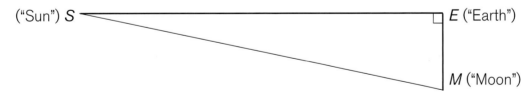

2. *a.* Compare your angle measurement with others in your group or your class, depending on your teacher's instructions, and devise a method for coming up with a single estimate of the angle. Give your estimate, and explain your method.

Navigating through Measurement in Grades 9–12

Angling for the Distance (continued)

Name _____

    b. Try to come up with an estimate of the error associated with your measure.

3. a. Record your collective estimate of θ (for m∠SEM) in the drawing. Again enter your own measurement of *EM*.

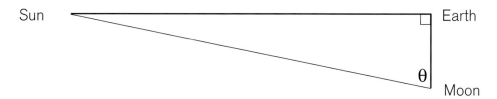

    b. Considering θ together with what you know about right triangle trigonometry, what can you conclude about the ratio of *EM/ES*? *Note:* If you haven't yet studied trigonometry, your teacher will supply some assistance.

    c. Using θ and your measurement of *EM*, find *ES*.

    d. Analyze the error in your estimate of *ES*.

    e. Check your results in 3(*d*) by measuring *ES* directly.

4. a. Aristarchus estimated m∠SEM as 87° when the moon was exactly at the half. Using this angle, what would he get as a value for *EM/ES*?

    b. If you use Aristarchus's value for *EM/ES* from 4(*a*) and take 239,000 miles as an estimate of the distance from the earth to the moon (*EM*), what do you get as an estimate of the distance from the earth to the sun (*ES*)?

Angling for the Distance (continued)

Name _____

c. Compare your value for *ES* from 4(*b*) with the earth-sun distance that NASA uses: 93 million miles.

5. a. If someone uses Aristarchus's method to estimate the ratio *EM/ES*, what are some of the possible sources of error to take into account?

b. The actual sun-earth-moon angle ($\angle SEM$) at the half-moon is close to $89°51'30''$. Use this angle measurement to estimate the percentage of error in Aristarchus's measurement of *EM/ES*.

6. a. If Aristarchus's value for $m\angle SEM$ had been only $(1/10)°$ greater than the quite accurate value of $89°51'30''$ given in 5(*b*), what would be the relative error of measurement in the value that Aristarchus would have computed for *EM/ES*?

b. What does this imply about Artistarchus' method?

c. Explain why in some cases a small error in a measurement can lead to a large relative error of measurement.

Starbucks Expansion

Name _____

For this activity, you will need the number of stores that Starbucks has operated each year since the company opened its first store. You can obtain this information from the Internet. One suitable site is the following:

- http://www.starbucks.com/aboutus/timeline.asp

Make an electronic file of your data, or record them with pencil and paper.

1. a. With the other members of your group, discuss the source(s) of the data and the time(s) of year when the counts were taken, if this information is available.

    b. If different members of your group obtained different data, decide on one set of numbers that everyone in the group will use.

    c. Discuss how reliable you think the data are.

2. a. On your calculator or an electronic spreadsheet, enter the years in **L1** (list 1) and the number of stores in **L2** (list 2). Use the numbers 0, 1, 2, …, to record the years. For example, let 0 represent the year Starbucks opened its first store and 1 represent the next year.

    b. Create a scatterplot showing the years against the number of stores.

3. Consider the shape of the plot:

    a. Does the number of stores increase or decrease over time?

    b. What would the graph look like if the number of stores increased at a constant rate?

    c. From your graph, would you say that the rate of growth in the number of stores is increasing, or is it decreasing?

    d. How many Starbucks stores do you estimate there will be at the end of the next year after the last one for which you have data?

4. a. Find the line that is the best fit for your data set, and show its graph on your scatterplot of the data.

    b. Is this line a reasonable fit for the plotted data points?

    c. From the information that you have gathered, graphed, and computed so far, would you say that the data are linear?

Starbucks Expansion (continued)

Name _____

d. If you were an investor, would you be willing to use this line of best fit to predict the growth of the company over the next several years?

5. If the data were exponential, then the relation between the number of stores (s) and the years (y) could be expressed as $s(y) = Ab^y$, where A is the initial value (a positive number) of s at time 0, and b is a constant: the ratio of consecutive terms in the sequence $Ab^0, Ab^1, Ab^2, \ldots$. For example,

$$\frac{s(3)}{s(2)} = \frac{Ab^3}{Ab^2} = b.$$

a. Compute the ratios of consecutive terms for the first few terms in the list **L2**.

b. Are the ratios approximately constant? _____

c. Do the data appear to be exponential? _____

6. a. Find the exponential function that is the best fit for your data, and show its graph on your scatter plot of the data.

b. Write the function in the form $s(x) = Ab^x$.

c. From the information that you have collected, graphed, and computed so far, would you say that the growth of the number of stores is linear, or is it exponential?

d. Which model do you think that the company should use to predict the number of new stores that it can open in the next several years?

7. a. Determine how many stores Starbucks would need to add in the next year to maintain an exponential growth pattern.

b. With the exponential model, how long would it take for the company to reach 100,000 stores?

c. Do you think that the exponential model would be a suitable model for the next 20 years of the company's life?

d. Would the linear model be suitable?

Golf Ball Boogie

Name _____

In 1940, noted photographer Harold E. Edgerton used a strobe light to capture the motion of a bouncing golf ball in a multiple-exposure photograph, "Golf Ball Bounce." Using a copy of this remarkable photograph that your teacher shows you or helps you find on the Internet, you will make measurements, scale them to reflect the actual situation that the photograph depicts, and use your measurements to develop a model of the golf ball's motion.

1. Look closely at the photograph. Determine the position of the surface on which the golf ball is bouncing.

2. a. In which direction is the ball moving in the photograph—from left to right or from right to left? _____ How do you know?

   b. Focus on the "middle" bounce in the photograph—that is, the segment of the ball's path from the ball's first contact with the surface to its next contact. Decide how to measure the ball's distance from the surface at each instant during this bounce. Will you measure to the top, bottom, side, or center of each ball-image?

   c. Which ball image in this middle bounce corresponds to the first instants in the rebound of the ball?

3. Continuing to examine the "middle" bounce, use your centimeter ruler to measure the vertical distance of each image of the ball to the surface. Start with the first image that shows the ball going up, and progress to the last image that shows the ball going down. Record your results in the data table:

Image	0	1	2	3	4	5	6	7	8	9	10	11	12	13	14	15	16	17	18
Distance (cm)																			

4. Assume that the strobe light flashed every 3 hundredths of a second. About how long was the ball in the air during the middle bounce?

5. a. Your teacher has given your group an actual golf ball along with a tape measure. Measure the circumference of the golf ball. What is its diameter in centimeters?

   b. Measure the diameter of the golf ball's image in the photograph. What is its diameter in centimeters?

   c. By what scale factor would you have to multiply the image's diameter to estimate the actual golf ball's diameter? Be careful with your units.

Golf Ball Boogie (continued)

Name _____

6. In centimeters, estimate the golf ball's actual maximum distance from the surface on which it was bouncing? *Hint:* Use the scale factor that you found in 5(c).

7. Convert all the distance measurements that you made from the photograph and recorded in the table in step 3 to estimates of the actual vertical distances of the ball from the surface. Record your values in the following data table:

Image	0	1	2	3	4	5	6	7	8	9	10	11	12	13	14	15	16	17	18
Time (sec.)	0	.03																	
Distance (cm)																			

8. Using a calculator or an electronic spreadsheet, enter the times (seconds) and the vertical distances (centimeters) from your table in step 7 into lists **L1** and **L2**, respectively.

9. a. Using an appropriate viewing window (a window in which all points are visible) of your calculator or the graphing capabilities of your spreadsheet software, make a scatterplot of your data.

   b. Sketch your plot from 9(a) on the axes shown, labeling them appropriately and giving the distance from the origin to the first tick mark in both the horizontal and vertical directions.

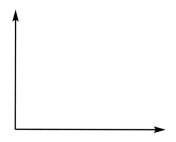

10. Thanks to Galileo, we know that a free-falling object's vertical distance from the surface of the earth is given by a quadratic equation whose first coefficient is 4.9 if the distance is in meters and the time is in seconds.

Golf Ball Boogie (continued)

Name _____

a. Find a quadratic polynomial that fits your data. *Hint:* Use 490 as your leading coefficient, since your data are in centimeters, and try different values for *a* and *b* in the equation $d(t) = 490t^2 + a \cdot t + b$.

b. Does your quadratic model do a good job of accounting for the data from the photograph?

c. What are the units of distance and time for the coefficients *a*, *b*, and 490?

Bouncing Ball

Name _____

In this exploration, you will make measurements and analyze the bouncing of an actual ball. For these tasks, the members of your group will share a ball, an electronic motion detector, a graphing calculator, and a cable to connect the calculator to the motion detector. The steps in the process are written for a Texas Instruments Ranger motion detector and a TI-83 graphing calculator, but your teacher can help you adapt the instructions to other devices, if necessary.

1. a. Connect the Ranger motion detector to the shared TI-83 calculator with the cable. Press **82/83** on the inside of the Ranger to load the Ranger programs into the TI-83.

   b. Press **PRGM** and **1** to place **prgmRANGER** on the calculator's home screen. Press **ENTER** to run the Ranger program. Press **ENTER** again at the title screen.

   c. Press **3** to select **APPLICATIONS,** and then press **2** to select **FEET** as the units for your measurements. Press **3** to select the application **BALL BOUNCE,** and press **ENTER** to prepare the calculator to accept data.

2. a. Select a student in your group to hold the Ranger at a height of about 6 feet from the floor, with the circular screen facing down. This student will press the **Trigger** button on the Ranger when everything is set. Select a second student to hold the ball at about 4 feet from the floor, directly under the motion detector's screen. Choose a third student to hold the connected TI-83 out of the way of the ball. Choose a fourth student to give a signal for the experiment to begin.

   b. At the signal, the first student should press the **Trigger** button on the Ranger. The second student should then let go of the ball. (Don't be discouraged if your group needs to repeat this process several times to obtain good data.)

3. The Ranger will collect distance and time data for 4 seconds. (The 4-second period is preset by the Ranger program.). The calculator will then display the data. The times are stored in **L1**, and the distances are stored in **L2**, on the calculator. The display should look something like the following:

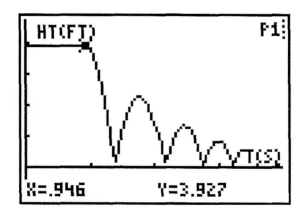

Bouncing Ball (continued)

Name _____

4. Move the cursor along your graph and answer the following questions:

   a. From what height did the ball originally drop? _____

   b. What were the times of the ball's first five bounces? (Give the times when the ball struck the floor.) _____, _____, _____, _____, _____

   c. What was the height of the ball's first rebound? _____ Its second? _____ Its third? _____ Its fourth? _____

5. Using the **Link** feature of the TI-83, send lists **L1** through **L6**, which contain the necessary data for the Ranger program, to each of the students in your group.

6. Have each student in your group select a different one of the parabolic shapes in the graph to analyze separately on his or her calculator. You can make the selection with the **Select Domain** option of the **Plot Tools** menu in the Ranger program:

   a. Graph **DIST–TIME**, and press **ENTER**.

   b. Press **1** for **Select Domain.**

   c. Move the cursor to the first point of your parabola, and press **ENTER**. Then move the cursor to the last point in your parabola, and press **ENTER** again. (Select points that are clearly above the *x*-axis to be sure that your points are not on someone else's parabola.)

7. Use the **STAT CALC QuadReg** command to approximate your data with a parabola. On the home screen, the command should look like **QuadReg L1,L2,Y1.** This command will store the quadratic function that best fits your data (that is, your distances and times) to **Y1** on the **Y=** menu.

8. Plot your data and graph your function. Do these tasks with the **Plot** feature of the TI-83:

   a. Press **2nd STAT PLOT,** and then press **ENTER.**

   b. Select **ON** and **boxes**, and enter **L1** in **Xlist** and **L2** in **Ylist**.

   c. Press **ZOOM ZoomData**, and then press **GRAPH** to plot the data and graph the function in **Y1** of the **Y=** menu.

9. a. Copy the graph from your graphing window.

   b. Label the horizontal and vertical axes, place tick marks on the axes, and label one of the tick marks on the horizontal axis and one on the vertical axis.

   c. Identify the maximum point of the data plot and the maximum point of the function graph.

Bouncing Ball (continued)

Name _____

10. Compare your data plot and graph with those of the other members of your group. Also compare your expression for the **Y1** function with theirs.

    a. Why do the graphs fill the screen now when originally they occupied only a small portion of it?

    b. Do the graphs look the same in the graphing windows? _____ How can this be so when the maximum values of the bounces are different?

    c. How do the values of the leading coefficient of the functions in **Y1** compare?

    d. How do the leading coefficients influence the shapes of the graphs?

11. Suppose that a certain ball rebounds to 2/3 of its previous rebound height with each bounce. Prove that the rebound height that the ball reaches between bounces is given by

    $$h_n = \left(\frac{2}{3}\right)^n h_0,$$

    where h_n is the height of the nth rebound, and h_0 is the initial height from which the ball was dropped.

12. a. Using the initial height of your group's ball as h_0, evaluate the function in step 11 as a model for your data. Is it a good model?

    b. Could you improve it as a model by using a fraction other than 2/3?

    c. If so, what fraction would you use?

13. When the ratio of consecutive terms of a sequence is constant, as with the consecutive rebound heights in step 11, the sequence is *geometric.*

    a. If the rebound heights of your ball form a geometric sequence, should your ball still be bouncing? _____

    b. Why, or why not?

Most Like It Hot

Name _____

In this exploration, you will make measurements and analyze the cooling of a cup of hot coffee over a short period of time. For these tasks, the members of your group will share an electronic temperature probe, a data collection device that works with the probe, a graphing calculator, and a cable that can connect the data collection device to the calculator. The steps in the process are written for a Texas Instruments CBL 2 data collection device and a TI-83 Plus graphing calculator, but your teacher can help you adapt the instructions to other devices, if necessary.

1. a. Connect the CBL 2 data collection device and the TI-83 Plus graphing calculator with the cable.

   b. Connect the temperature probe to the CBL 2 by plugging the probe into the slot on the CBL 2 marked **Ch1**.

   c. Load the **DATAMATE** program into the TI-83 by pressing the **TRANSFER** button on the CBL 2. Press the **APPS** button on the calculator, and use the down arrow to select the **DATAMATE** program.

   d. Press **ENTER** to start the program.

   e. Press **1** to set up the time graph mode to collect 20 samples at 1-minute intervals.

2. a. Place the temperature probe in a cup of hot coffee (at least 150° Fahrenheit, or 66° Celsius).

   b. Wait 15 seconds, and press **2** on the calculator to start collecting temperature data. The Datamate program will record the temperature of the probe every minute for 20 minutes.

3. When the program stops collecting data, look at the graph of the data on your calculator. Make a copy of the displayed graph on the axes provided. Be sure to show the tick marks on your graph.

4. a. Press **ENTER,** select **TOOLS**, and press **2** to store or retrieve the data in your calculator. The calculator will store the time and temperature data in lists **L1** and **L2**, respectively.

   b. Quit the Datamate program, and examine the data by pressing first **STAT** and then **ENTER** to edit or view the lists.

Most Like It Hot (continued)

Name _____

5. Refer to the data as you answer the following questions:

   a. What are the units of the time measurements? _____ Show these units and the number of units between tick marks on the graph that you made in step 3.

   b. What are the units of the temperature measurements? _____ Show these units and the number of units between tick marks on your graph from step 3.

   c. Is the temperature increasing or decreasing as the time increases?

   d. Describe the rate at which the temperature changes as the time increases.

   e. From the graph that you copied from your calculator, would you say that the relationship between the time and the temperature is linear? _____ Why, or why not?

   f. Would you expect the temperature of the coffee in the cup to drop below the temperature of the room? _____ Why, or why not?

6. a. Use the CBL 2 and the calculator to determine the ambient temperature of the room. For this task, hold the temperature probe in the air while you run the Datamate program. The temperature of the room will appear in the upper right corner of the screen. What is the ambient temperature of your classroom? _____

   b. On your graph from step 3, draw a line representing the room temperature for the time period of the experiment.

   c. Consider the following function types:

   - Linear
   - Quadratic
   - Cubic
   - Exponential
   - Logarithmic
   - Rational

   Which type of function do you think would give the best model of the trends in your coffee temperatures?

Most Like It Hot (continued)

Name _____

    d. What features does your graph share with this type of function?

7. *a.* Find a linear function that approximates the data that you have collected.

    b. Use the curve-fitting feature in the **STAT** menu of your calculator to fit a linear function to your data.

    c. Compare your linear function with the one that your calculator computed. Which one fits the data better?

8. *a.* Find an exponential function of the form $y = a \cdot b^x + c$ that approximates the data that you have collected. (*Hint:* Let *c* be the ambient temperature.)

    b. Use the curve-fitting feature in the **STAT** menu of your calculator to fit an exponential function to your data.

    c. Compare your exponential function with the one that your calculator computed. Which one fits the data better?

9. *a.* Which of the models that you constructed in steps 7 and 8 appears to be the best model of the cooling phenomenon? _____

    b. Justify your answer by considering the differences between consecutive temperatures in the data set.

Solutions for the Blackline Masters

Solutions for "Counting on Commensurability"

Quest for the Golden Ruler–Part 1

1. *a.* 271 jarboos.
   b. 300 jarboos.
2. *a.* Since neither 11 nor 7 divides 360 evenly, the lengths of the segments would not be whole numbers of jarboos.
   b. 77 units (the least common multiple of 7 and 11).
3. *a.* Yes.
   b. The smallest such number would be the least common multiple of 3798 and 482 (which, for the record, is 915,318). In fact, we could (in theory) divide the ruler successfully no matter what whole numbers appeared in the denominators of these two fractions.

Solutions for "That's Irrational"

Quest for the Golden Ruler–Part 2

1. *a.* n enths.
   b. By the Pythagorean theorem, the length of the hypotenuse would be
   $$\sqrt{n^2 + n^2} = \sqrt{2n^2} = n\sqrt{2} \text{ enths.}$$
2. If the hypotenuse has a length of m enths, then the result from step 1 shows that $n\sqrt{2} = m$. Dividing both sides of this equation by n, we see that
   $$\sqrt{2} = \frac{m}{n}.$$
3. $\left(\frac{m}{n}\right)^2 = 2 \Rightarrow \frac{m^2}{n^2} = 2 \Rightarrow m^2 = 2n^2$.
4. Since m^2 is twice an integer, m^2 must be even. If m were odd, then m^2 would be odd. Thus, m is even and a multiple of 2. Therefore, m^2 is a multiple of 4.
5. If m^2 is a multiple of 4, then $m^2/2 = n^2$ must be a multiple of 2. Now, if n were odd, then n^2 would be odd and thus not a multiple of 2. Therefore, n is even.
6. Since m/n is in lowest terms, both m and n can't be even. Therefore, since m is even, n must be odd.
7. *a.* The conclusions in steps 5 and 6 are mutually contradictory and hence cannot both be true.
   b. We conclude that the assumption in step 3, which led to both of these mutually contradictory conclusions, cannot be true. No matter what units we use to measure the leg of this triangle, the hypotenuse cannot be measured as a whole number of those units.

Solutions for "Super Bowl Shipment"

Approximately Speaking–Part 1

1. *a.* $100 \times (150 \times (8.4 \div 16) + 1) = 7{,}975$ lbs.
   b. Yes, the total weight is less than 8,000 lbs.
   c. The probability is essentially zero.

2. *a.* The range of possible weights of a doll is 8.35 to 8.44 oz.
   b. The largest weight of a doll is 8.45 oz.

3. *a.* The range of possible weights of the packing material is 0.5 to 1.4 lbs.
   b. The most that the packing material might weigh is 1.5 lbs.

4. *a.* $100 \times (150 \times (8.45 \div 16) + 1.5) \approx 8{,}072$ lbs.
   b. No, the total weight is not under the 8,000-pound limit.
   c. Students' answers will vary. On the one hand, if the driver feared that the extra 100 pounds would damage his truck, he might want to be cautious and use the higher estimate. On the other hand, if he regarded the road inspection as his only worry, he might be willing to use the lower estimate and take the risk, especially since at worst he would be only slightly overweight, and the paperwork showing the legal estimate might be sufficient to get him off the hook.

Solutions for "Paula's Popcorn Box"

Approximately Speaking—Part 2

1. $25 \times 17 \times 8 = 3{,}400$ cm^3.

2. Yes, because the dimensions would all "round down." (The rounding in this activity assumes the usual convention for rounding numbers taught in most high school mathematics classes. Scientists, however, sometimes use a different method, summed up by "When in doubt, even it out." This method would round 24.5 to 24 and 7.5 to 8, for example, making the last digits in the rounded numbers even.)

3. *a.* $25.4 \times 17.4 \times 8.4 \approx 3{,}712.5$ cm^3.
   b. This volume exceeds the volume in step 1 by about 312.5 cm^3.

4. Yes, because these numbers would all "round up."

5. *a.* $25.4 \times 16.5 \times 7.5 \approx 3{,}032$ cm^3.
   b. This volume falls short of the volume computed in step 1 by about 368 cm^3.

6. *a.* The respective errors amount to about 52 popped kernels over and 61 popped kernels under.
   b. Paula might round to the nearest thousand and report the volume of her box as 3000 cm^3, thus giving one significant digit (since the product involved a factor with one significant digit). Or she might report the volume of her box as 3400 ± 400 cm^3 to emphasize the uncertainty interval or the range of possible values. (See the discussion of this problem in "About This Book," pp. xi–xii.)

Solutions for "Rounding Numbers in a Sum"

Approximately Speaking—Part 3 (Class Discussion)

1. *a.* No, if each person in a group of 21 people weighed over 240 lbs., the whole group could not ride together safely in the elevator; $21 \times 240 = 5{,}040 > 5{,}000$.
   b. The engineer doesn't worry because he realizes that it is highly likely that a typical group of 21 people will have some below the average weight and some above the average weight, with the average being about 175 lbs. Furthermore, it is very difficult to fit 21 people into most elevators. It is hard to conceive of any group of 21 people with each weighing over 240 lbs. wanting to attempt such a feat.

2. *a* and *b.* Since 175 lbs. is about average, the flight engineer can simply multiply 64 by 175 to get an estimate of the total weight of the 64 people. People who weigh a little more or less will essentially cancel each other out, except for very small and very large individuals, who are accounted for in the separate count.

3. *a* and *b*. In the long run, the amounts rounded up and the amounts rounded down should essentially cancel each other out, leaving the dollar total as a pretty good estimate of the actual total in dollars and cents.

Solutions for "The Right Rope"

Early Measuring Devices—Part 1

1. *a.* Ancient engineers could shape the rope into a 3-4-5 triangle, which (by the Pythagorean theorem) is a right triangle.
   b. The ancient engineers would have been most interested in the triangle's right angle.

2. *a.* Most modern carpenters use a sturdy metal "carpenter's square" to measure right angles.
   b. The carpenter's square would be more reliable than the rope. It can be manufactured with considerable accuracy and will hold its shape.

Solutions for "Why Ships Measure Speed in Knots"

Early Measuring Devices—Part 2

1. The glass container full of sand was an hourglass (a chronometer), a very effective primitive device for measuring time. It worked by allowing a fixed amount of sand to flow from an upper chamber through a small opening to a lower chamber. The time unit was controlled by the amount of sand in the hourglass. The crew needed the hourglass to know when to stop feeding out rope.

2. The number of knots measured the distance traveled by the ship in the fixed period of time measured by the hourglass. Distance divided by time gave the rate of the ship.

3. Given the other information that a captain had relating to location, we can probably regard the method as valid for most trips. In our era of motorized highway travel, times of arrival can often be estimated to the nearest quarter hour (consider a bus schedule). By contrast, in the age of the hourglass, times of arrival were probably estimated to the nearest day or even week, depending on the length of the journey. Speed as measured by knots would probably be valid for such estimates. Of course, weather and sea conditions could affect the validity of the measures.

4. *a.* Generally speaking, the method would be reliable, since the sailor who was doing the measuring probably rounded to the nearest knot.
   b. Under ordinary conditions, two different members of the crew would get the same readings.
   c. Conditions at sea would affect the consistency of such measurements. Calm waters would give more consistent measurements than tempestuous seas would.

5. *a.* $\text{Speed} = \dfrac{\text{distance}}{\text{time}} = \dfrac{2 \times 47.25 \text{ feet}}{28 \text{ seconds}} = 3.375 \text{ ft/sec.}$

   b. $\dfrac{3.375 \text{ feet}}{1 \text{ sec}} \times \dfrac{1 \text{ mile}}{5280 \text{ feet}} \times \dfrac{3600 \text{ sec}}{1 \text{ hr}} = 2.3 \text{ mph.}$

6. $\dfrac{2.3 \text{ mi}}{1 \text{ hr}} \times \dfrac{1 \text{ nautical mi}}{1.15 \text{ mi}} = 2 \text{ nautical mph} = 2 \text{ knots.}$

7. A speed measured by 2 knots in the rope came to be known as 2 "knots." The measures of 47.25 feet and 28 seconds were the actual measures used by the sailors, and that is why to this day 1.15 mph = 1 nautical mph, or 1 knot.

Solutions for "More Measurement Methods"

Early Measuring Devices—Part 3

1. You could tie a heavy knot in one end of the rope and drop it down the well until you felt the rope go slack. The rope would then have reached the bottom of the well. You could then mark the top of the rope at the edge of the well, pull up the rope, and use your ruler to measure the distance from the knot to the mark on the rope. (Of course, this method would work only if your rope was longer than the well was deep, and if the knot didn't float.)

2. You could stick the pencil perpendicularly into the ground and then use the ruler to measure the length of the exposed pencil (p) and the length of its shadow (s). Next, you could use the ruler again, this time to measure the length of your stride. Then you could pace out the length of the shadow of the building (b) and use your stride length to convert your shadow measurement to feet. If x is the height of the building, you could find x by computing the following proportion:

$$\frac{x}{b} = \frac{p}{s}.$$

3. A flash of lightning travels to the eye at the speed of light (186,000 miles per second, which is virtually instantaneously), but the accompanying clap of thunder would travel to the ear at the much slower speed of sound (approximately 1 mile every 5 seconds, or about 1100 ft/sec). Therefore, you could use your watch to count the number of seconds between the lightning and the thunder and divide it by 5 to get the approximate number of miles between yourself and the storm. Expect a variety of answers to this question. After students have had a chance to grapple with the problem, you might give them the "1 mile in 5 seconds" rule.

4. You could use the pencil to mark the beginning and end of your stride on the ground, and then you could use the index card to measure the length of your stride in inches. Next, you could pace out the length and width of the soccer field. Then you could convert paces to inches and inches to feet. Finally, you could use the formula $A = l \times w$ to compute the approximate area of the field in square feet.

5. The length of the femur (f) is a linear measure, as is the height (h) of the dinosaur. Therefore, you could assume that the ratio (k) of a dinosaur's height to the length of its femur would be approximately constant in each species of dinosaur. You could check paleontologists' records to find the value of k for the particular species of dinosaur (assuming that an intact skeleton had been found at some time in the past). You could then use the formula $h = k \times f$ to find the height of the dinosaur.

6. You could lay out pennies and nickels on the 11-inch edge of the sheet of paper, working to cover the edge exactly with whole numbers of coins. You could do this with 6 nickels and 8 pennies. You could cover the 8-and-one-half-inch edge of your paper in the same way with 3 nickels and 8 pennies. If you saw that 8 penny-diameters + 6 nickel-diameters = 11 inches, and 8 penny-diameters + 3 nickel-diameters = 8.5 inches, then you could deduce that three nickel diameters must equal 2 1/2 inches. This would tell you that one nickel is 5/6 of an inch in diameter. You could then see that 8 penny-diameters must equal 6 inches, which means that a penny is 3/4 of an inch in diameter. You could then use the fact that the circumference of a circle is equal to π times the diameter to determine that the circumference of a penny is $3\pi/4$ inches, and the circumference of a nickel $5\pi/6$ inches. (To check their answers, students can lay out 6 nickels on a ruler and see that they cover 5 inches, and they can lay out 4 pennies on a ruler and see that they cover 3 inches.)

Solutions for "Mathematical Goat"

1. a. The new region is shown.

    b. The original fenced region was a square with 12-foot sides. The new grazing region outside the pen is three-quarters of a circle with a radius of 8 feet.

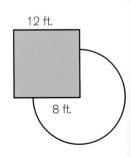

2. a. The formula for the area of a circle: $A = \pi r^2$.
   b. The original square has an area of 144 ft²: $A = s^2$, so $A = 12^2 = 144$ ft².
      The new area is 48π, or just over 150 ft²: $A = (3/4)\pi r^2$, so $A = (3/4)\pi 8^2 = 48\pi \approx 150.794$ ft².
   c. The two regions, though different in shape, are not so very different in area.

3. a. Students' answers will vary; an approximate count is 153 units, with each unit representing 1 square foot of area.
   b. A count of approximately 153 units supports the computed answer for the new area, as given above.

4. This new grazing region consists of three-quarters of a circle with a radius of 14 feet, plus two quarter-circles with radii of 2 feet, as shown.

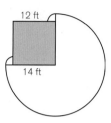

5. a. The formula for the area of a circle: $A = \pi r^2$.
   b. The new area is 149π, or about 468 ft²: $A = (3/4)\pi 14^2 + (1/2)\pi 2^2 = 149\pi \approx 468.089$ ft².

6. a. Students' answers will vary; an approximate count is 471 units, with each unit representing 1 square foot of area.
   b. A count of approximately 471 units supports the computed answer for the new area, as given above.

7. The new grazing area defined by the 14-foot rope is about 3 times as large as the region defined by the 8-foot rope.

8. a. The new grazing region consists of four congruent sectors of radius 14 feet, plus four congruent isosceles triangles, as shown.
   b. An approximate count of 1276 units gives an estimated total area of about 1276 square feet.
   c. The grazing region is geometrically complex, and students would need trigonometry to compute its area accurately. However, with a protractor, they can compute a good estimate. They can use the Pythagorean theorem to find the height of each isosceles triangle:

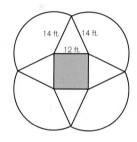

$$h^2 = 14^2 - 6^2 = 196 - 36 = 160$$

$$h = \sqrt{160} \approx 12.65 \text{ ft.}$$

For a triangle, $A = (1/2)bh$, so students can find the area of each isosceles triangle: $A \approx (1/2)(12)(12.65) \approx 76$ ft². The base angles of these triangles have a cosine of 6/14, and hence, they measure about 64.62°, since arccos 6/14 ≈ 64.62. Students can now determine the measure of the angle in each sector, since each of these angles is 270° less two of the base angles, or about 140.75°. Thus, they can calculate the area of each of the four sectors as about 241 ft²: $(140.75/360)\pi 14^2 \approx 240.76$ ft². So the total grazing area, consisting of the four isosceles triangles and the four sectors, is about 1267 ft².

9. a and b. The 12-foot sides of the square pen would have less and less effect on the shape of the grazing area as the length of the rope increased. With a rope 1 mile long, the grazing area would approximate a circle with a radius of 1 mile.

Solutions for "Chip off the Old Block"

1. a. $V_{sphere} = (4/3)\pi r^3$.
   b. $SA_{sphere} = 4\pi r^2$.

2. The greatest radius that the woodcarver can achieve is 3 centimeters.

3. a. $V = (4/3)\pi r^3$. Therefore, letting $r = 3$, we get $V = 36\pi$ cm³. Using $\pi \approx 3.14$ gives us a volume of about 113 cm³.

b. This volume is the maximum because it uses the greatest possible radius.

4. a. $V_{cylinder} = \pi r^2 h$.
   b. $SA_{cylinder} = 2\pi r^2 + 2\pi r h$.

5. a. The greatest radius that the woodcarver can obtain for the cylinder is 3 centimeters.
   b. The greatest height that the woodcarver can obtain for the cylinder is 6 centimeters.

6. a. $V_{cylinder} = \pi r^2 h$. Thus, letting $r = 3$ and $h = 6$, we get $V = 54\pi$ cm³. This volume is about 170 cm³.
   b. Yes, this volume appears to be the maximum. Although the woodcarver could cut cylinders with larger bases from the block, their heights would be small.

7. The volume of each cube is $6 \times 6 \times 6$, or 216, cm³.

8. a. Approximately 113/216, or about 52 percent. The volume of the sphere is thus just over one-half the volume of the cube.
   b. Approximately 170/216, or about 79 percent. The volume of the cylinder is thus just over three-fourths that of the cube.

9. a. The fraction of the cube that is discarded for the sphere is approximately 103/216, or about 1/2 of the cube.
   b. The fraction of the cube that is discarded for the cylinder is approximately 46/216, or about 1/4 of the cube. So the woodcarver wastes about half of one of his cubes when he cuts out the sphere and about a quarter of his other cube when he cuts out the cylinder.

10. $54\pi/36\pi = 3/2 = 1.5$. The volume of the cylinder is exactly 150 percent of that of the sphere.

11. a. Since the largest radius is 3, the surface area of the largest sphere is $4\pi(3^2)$, or 36π, or about 113 cm².
    b. The surface area of the largest cylinder is $2\pi(3^2) + 2\pi(3)6$, or 54π cm².
    c. The ratio is 54/36, or 3/2, or exactly 1.5.

12. a. They are the same; both are 1.5.
    b. This result is not typical. To be sure of this, students need only calculate the dimensions of a cylinder whose radius and height are different from the one in the activity but whose volume is still 150 percent of the volume of the sphere in the activity. The surface area of the cylinder will not be 150 percent of the surface area of the sphere.
    c. There are in fact many cylinders of various dimensions that have exactly 150 percent of the volume of the sphere in this activity. But the surface area of only one of those cylinders is exactly 150 percent of the surface area of the sphere. That unique cylinder is the cylinder in this activity. Your advanced students should be able to prove this algebraically. Archimedes deserved to be excited.

13. a. The results would be the same.
    b. In general, when a cube has a side of length *a*, then the largest sphere and largest cylinder that can be cut from such a cube both will have radii of *a*/2, and the cylinder will have a height of *a*. Students can use basic algebra and the formulas for volume and surface area to generalize their results.

Solutions for "A Base on a Face"

Cones from Cubes—Part 1

1. $V_{cone} = 1/3 \pi r^2 h$.

2. a. The greatest radius that the woodcarver can give the cone is 3 centimeters.
   b. The greatest height that the woodcarver can give the cone is 6 centimeters.

3. a. $V_{cone} = 1/3 \pi r^2 h$. Therefore, letting $r = 3$ and $h = 6$, we get $V = 18\pi$ cm³.
   b. Using $\pi \approx 3.14$ gives a volume of about 57 cm³.

a. The volume of the cube that the woodcarver had at the beginning is $6 \times 6 \times 6$, or 216, cm³.

b. The volume of the cone to the volume of the cube is

$$\frac{57}{216} \approx .263,$$

or about 26 percent. Thus, the volume of the cone is just over a quarter of the volume of the cube.

a. Students should answer no, since this cone does not have the greatest volume of all possible cones that the woodcarver could cut from the cube.

b. While locating the base of the cone on a face of the cube may seem intuitively to be the obvious position for such a cone, it does not give a cone of maximum volume from the cube. A cone of maximum volume is obtained by inscribing the base of a cone in a regular hexagon formed by the slicing of a plane through the cube.

Solutions for "Going for the Max"

Cones from Cubes—Part 2

One possible solution is shown on the activity sheet. Each vertex of any such regular hexagon is the midpoint of one edge of each of the six faces of the cube. Four different hexagons are possible, but they are all congruent.

a. The edge *QR* of the cube is 6 cm. Using the Pythagorean theorem, we can calculate the face diagonal *PQ* of the cube as $PQ^2 = 6^2 + 6^2 = 72$ cm². Thus, $PQ = \sqrt{72}$ cm. Again applying the Pythagorean theorem, we can calculate the main diagonal *PR* of the cube as $PR^2 = PQ^2 + QR^2 = 72 + 36 = 108$ cm². Thus, $PR = \sqrt{108}$ cm.

b. $PR \approx 10.4$ cm.

a. The plane of the hexagon cuts the cube into two congruent halves, bisecting the main diagonal *PR*. (See the illustration in number 2 on the activity sheet.) Thus, the height of the woodcarver's new cone is half the length of the main diagonal of the cube, or $h = (1/2)(\sqrt{108})$, or $3\sqrt{3}$ cm.

b. The height of the new cube is approximately 5.2 cm.

a. A side of the hexagon connects the midpoints of two adjacent sides of the face of the cube. (See the illustration on activity sheet.) Thus, by considering similar triangles, we can see that the length of this and every side of the regular hexagon must be half the length of a face diagonal. The side of the regular hexagon is

$$\frac{1}{2}(6\sqrt{2}), \text{ or } 3\sqrt{2}, \text{ cm.}$$

b. By using the Pythagorean theorem, we can find the altitude shown of the equilateral triangle in the hexagon. This is the apothem of the regular hexagon and is also the radius of the cone. We can compute this radius as follows:

$$r^2 + \left(\frac{3\sqrt{2}}{2}\right)^2 = \left(3\sqrt{2}\right)^2.$$

Thus, solving for *r*, we get $r = \sqrt{\frac{27}{2}} \approx 3.67$ cm.

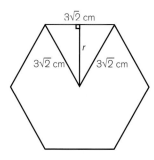

Radius of the cone: $r = \frac{3\sqrt{6}}{2}$

a. We can substitute for *h* and *r* in the formula to find the volume of the new cone.

$$V = \frac{1}{3}\pi r^2 h = \frac{1}{3}\pi \left(\sqrt{\frac{27}{2}}\right)^2 3\sqrt{3} = \frac{27\sqrt{3}\pi}{2} \text{ cm}^3$$

b. The volume of the new cone is about 73 cm³.

Navigating through Measurement in Grades 9–12

6. *a.* The volume of the original 6-centimeter cube was 216 cm³. The volume of the new cone compared to the volume of the cube is about 73/216, or about 34 percent of the volume of the cube. Thus, the volume of the new cone is very close to one-third of that of the cube.
   b. The volume of the cone whose base was on a face of the cube was about 26 percent of the volume of the cube. Thus, the new cone, with a volume that is about 34 percent of the volume of the cube, is substantially larger than the cone whose base was on a face of the cube.
   c. The volume of the new cone appears to be the maximum, since the base of this cone is the maximum for a cone inscribed in the cube. There are cones with heights that are greater than that of the new cone, but we know that those cones have smaller volumes, since the key determinant of a cone's volume is the radius of its base, because it is the radius that is squared in the formula for the volume of a cone.

Solutions for "Iterating on a Plane"

Measuring a Geometric Iteration–Part 1

1. *a.* The area at stage 0 is 9 in².
   b. The area at stage 1 is 5 in².

2. The area at stage 1 is 5/9 of the area at stage 0.

3. *a.* The perimeter at stage 0 is 12 in.
   b. The perimeter at stage 1 is 20 in.

4. From stage 0 to stage 1, the perimeter increases.

5. The stage 2 figure will have 25 squares.

6. *a.* At stage 2, the area of each of the squares is 1/9 in².
   b. At stage 2, the perimeter of each of the squares is 4/3 in².

7. The completed data table shows the values:

Stage	0	1	2	3	4	n
Number of squares	1	5^1	5^2	5^3	5^4	5^n
Total area (in²)	9	$9(5/9)^1$	$9(5/9)^2$	$9(5/9)^3$	$9(5/9)^4$	$9(5/9)^n$
Total perimeter (in)	12	$12(5/3)^1$	$12(5/3)^2$	$12(5/3)^3$	$12(5/3)^4$	$12(5/3)^n$

Students may note that the factors of 5, 9, and 12 recur in the table entries. This observation is not necessary but can greatly facilitate the students' recognition of the constant multiplier in the corresponding geometric sequences.

8. *a.* The constant multipliers are 5, 5/9, and 5/3.
   b. See the *n*th term entries in the data table.

9. At stage 15, the total number of squares in the figure first exceeds 6.5 billion, the estimate of the world's population. At stage 14, the number of squares passes 6 billion ($5^{14} = 6,103,515,625$). But stage 15 ($5^{15} = 30,517,578,125$) is the first to contain more than 6.5 billion tiny squares.

10. The entries in the data table give the perimeter in inches. The number of inches in a mile is 12 × 5280, or 63,360. Stage 17 is the first to have a perimeter greater than 1 mile: $12(5/3)^{17} \approx 70,894$ inches.

Solutions for "Iterating in 3-D"

Measuring a Geometric Iteration–Part 2 (Assessment)

1. *a.* The surface area at stage 0 is 54 in².
   b. The surface area at stage 1 is 54 in².

2. a. The volume at stage 0 is 27 in^3.
    b. The volume at stage 1 is 9 in^3.

3. The completed data table shows the values:

Stage	0	1	2	3	4	n
Number of cubes	1	9^1	9^2	9^3	9^4	9^n
Total surface area (in^2)	54	54	54	54	54	54
Total volume (in^3)	27	$27(1/3)^1$	$27(1/3)^2$	$27(1/3)^3$	$27(1/3)^4$	$27(1/3)^n$

  a. The number of cubes is increasing and has a constant multiplier of 9.
  b. The surface area remains constant at 54 in^2.
  c. The volume is decreasing and has a constant multiplier of 1/3.

Solutions for "Building Pyramids with Cubes"

Discovering the Volume of a Pyramid—Part 1

1. The volume of the two-layered pyramid is 5 cubic units.
2. The volume of the corresponding cubic prism is 8 cubic units.
3. The values are shown in the data table.

Number of layers in the "pyramid" of cubes	Volume of the "pyramid" of cubes	Volume of the corresponding cubic prism	Volume of "pyramid" / Volume of cubic prism
1	1	1	1
2	5	8	$5/8 = .625$
3	14	27	$14/27 \approx .518$
4	30	64	$30/64 \approx .469$
5	55	125	$55/125 = .44$
6	91	216	$91/216 \approx .421$
7	140	343	$149/343 \approx .408$
8	204	512	$204/512 \approx .398$
9	285	729	$285/729 \approx .391$
10	385	1000	$385/1000 = .385$
n	Volume of the previous stage + n^2, or $1^2 + 2^2 + 3^2 + 4^2 + \ldots + n^2$, which equals $(1/6)(n)(n+1)(2n+1)$	n^3	$\left(\frac{1}{6}\right)\frac{(n)(n+1)(2n+1)}{n^3}$, or $\frac{1}{3} + \frac{1}{2n} + \frac{1}{6n^2}$, which decreases to $\frac{1}{3}$ as n increases.

  a. Accept any reasonable answer that your students offer at this point, such as, "The ratio is decreasing as the shapes get bigger." In fact, the ratio in the last column decreases monotonically to 1/3, or .33333…, as a limit.
  b. Again, accept any reasonable answer. Although the ratio in the last column decreases, it will never be smaller than 1/3, so it will never reach 0, for example. Let your students discuss their ideas, but do not give away the fact that the ratio approaches 1/3. As the the pyramids of cubes become larger, they become closer and closer approximations of actual, smooth-sided pyramids. The reason for this is that the size of the cubic blocks is unchanging while the pyramids are becoming larger, so the "bumps" that the edges of the blocks create become less and less significant as the process of building pyramids continues.

In discussions with your students, you can relate this sequence of cube-pyramids to a regular square pyramid whose height and sides of the base are all one meter. By appropriate scaling of the cubes, the nth stage cubic pyramid (when the side of each cube measures $1/n$ meters) corresponds to this regular square pyramid and the prism containing it. As *n* gets larger, the cubic pyramids become closer and closer approximations of regular square pyramids.

Solutions for "Probing Pyramids with Spreadsheets"

Discovering the Volume of a Pyramid–Part 2

1 and 2. The students' spreadsheets should look something like the sample.

Number of layers	Volume of "pyramid"	Volume of cubic prism	Volume of "pyramid" / Volume of cubic prism
1	1	1	
2	5	8	

3. The sample displays the formulas, but only numbers should show in the cells on the students' spreadsheets.

	A	B	C	D
1	Number of layers	Volume of "pyramid"	Volume of cubic prism	Volume of "pyramid" / Volume of cubic prism
2	1	1	1	=B2/C2
3	2	5	8	

4. Again, the sample displays the formulas, but only numbers should show on the students' spreadsheets.

	A	B	C	D
1	Number of layers	Volume of "pyramid"	Volume of cubic prism	Volume of "pyramid" / Volume of cubic prism
2	1	1	1	=B2/C2
3	2	5	8	=B3/C3

5. The sample displays the formulas.

	A	B	C	D
1	Number of layers	Volume of "pyramid"	Volume of cubic prism	Volume of "pyramid" / Volume of cubic prism
2	1	1	1	=B2/C2
3	2	5	8	=B3/C3
4	=A3+1	=B3+(A4)^2	=(A4)^3	=B4/C4

6. A few sample rows of a spreadsheet are shown.

Number of layers	Volume of "pyramid"	Volume of cubic prism	Volume of "pyramid" / Volume of cubic prism
196	2529086	7529536	0.335888692
197	2567895	7645373	0.335875699
198	2607099	7762392	0.335862837
199	2646700	7880599	0.335850105
200	2686700	8000000	0.3358375

7. The method that students use to create their graphs will vary depending on their spreadsheet software and computer platforms.
8. The numbers in the last column get closer and closer to 1/3. The ratio will always be greater than 1/3.
9. $V = (1/3)Bh$.

Solutions for "Making a Formula"

Measuring the Size of a Tree—Part 1

Answers will vary. The authors are indebted to ninth graders at Galileo Academy of Science and Technology in San Francisco, California, for permitting their responses to appear as sample solutions. The selection is meant to be neither exhaustive nor representative of the class.

1. Sample attributes of a tree follow.

height of tree	length of leaves
trunk of tree	length of branches
width of leaves	number of roots
spread of leaves	spread of roots
color of leaves	color of roots
number of leaves	number of bugs
number of branches	number of spider webs
spread of branches	length of shadow
color of branches	number of dead leaves
age of tree	circumference of tree

2. In the list of sample attributes above, the underlined items are attributes that relate to a tree's size.

3. A sample completed table follows:

Attribute	Variable	Unit of measure	How would you make the measurement?
Trunk	T	inches	Measure circumference with a measuring tape
Length of branches	B	inches	Measure some branches with yardstick and compute average
Length of leaves	L	inches	Measure lengths of some leaves and compute average
Height of tree	H	inches	Measure indirectly using the protractor tool

4. *a* and *b*. A sample formula and explanation follow:

$$\text{Our formula: } L \times (T \times H)/2 + (T \times H)/2$$

T is the circumference of the tree, *H* is the height of tree, and *L* is the average leaf length.

By multiplying *T* and *H*, you get the [surface] area of [the tree] from the trunk and straight up to the top of the tree. And we took half of that area and multiplied it by the length of [the tree's] leaves, which equaled the top half of the tree. Then we took the rest of the area and added this, since it represented the bottom half of the tree, which has no leaves.

Note: These students intended to use surface area as the size of the tree. They distinguished between the leafy "top half of the tree" and the bottom half, which had no leaves.

Solutions for "Using a Formula"

Measuring the Size of a Tree—Part 2

Students' answers will vary. Selected responses from ninth graders at Galileo Academy of Science and Technology in San Francisco, California, follow.

1. *a.* A sample description of two trees is given:

   Tree 1
   Our first tree is located behind the fences and a couple of feet away from the entrance to the other building. This tree also has green leaves, [and] cobwebs, [and] the trunk is a little bit skinnier than the first one. This tree has some thorns.

Tree 2

Our second tree is located next to the benches and the entrance to the beanery. This tree has some dead leaves including cobwebs. The tree is about 2 stories high and the trunk looks fat to me.

b. Tree 1 is __2__ times *bigger / smaller* than tree 2.

2. Sample measurements of two trees follow, along with explanations of the measuring methods:

   Measurement 1 (for the attribute *T*, or circumference)

   Tree 1
   Circumference = 25″

   Tree 2
   Circumference = 29″

   How did you make the measurement? We put the tape measure around the trunk at of the tree to see what the size was. We did the same measuring as tree 1 to figure out the circumference of tree 2.

   Measurement 2 (for the attribute *L*, or length of leaves)

   Tree 1
   Leaves = 3.6″

   Tree 2
   Leaves = 3.6″

   How did you make the measurement? We took three leaves and measured their length and averaged the three numbers. We got the same answer for the average [for both trees].

   Measurement 3 (for the attribute *H*, or height)

   Tree 1
   Height = 181.5″

   Tree 2
   Height = 169.8″

   How did you make the measurement? We used the protractor tool and our calculators [for both trees], but a different person held the protractor [for tree 2].

3. Sample calculations and commentary follow:

   a. Our formula was $Size = L \times (T \times H)/2 + (T \times H)/2$.
      Our final calculation of the sizes of our two trees follows:

      Tree 1
      H = 181.5″
      T = 25″   Size = 10,436.25
      L = 3.6″

      Tree 2
      H = 169.8″
      T = 29″   Size = 11,325.66
      L = 3.6″

   b. Tree 1 is __0.9__ times *bigger / smaller* than tree 2.

   c. This is not that close to our answer in 1(*b*), where we predicted that tree 1 was only half as big [as tree 2]. The fatter trunk ended up not mattering that much, and tree 1 turned out to be a little bit taller. At least that is what our protractor tool told us. The leaves didn't really matter because the trees [both] had the same kind of leaves.

   Note: In this sample, the students made all their length measurements in inches, but they did not give units for their final measurements. The class addressed this issue when the students presented their poster.

Solutions for "If the Earth Is Round, How Big Is It?"

1. a. Assuming that the sun's rays are parallel, we know that $TB \parallel OS$. Therefore, by applying the alternate interior angle theorem, we find that $m\angle SOA = m\angle TBA = 7.2°$. Since $\angle SOA$ is a central angle of a great circle, then the arc that it subtends on the great circle is 7.2/360, or 2 percent, of the circumference of the great circle. Since $0.02 \times C = 500$ miles, $C = 25{,}000$ miles.

   b. A circumference of 25,000 miles gives a polar diameter of 7,958 miles ($25000/\pi \approx 7958$). (The actual polar circumference of the earth is close to 24,860 miles, and the polar diameter is 7,912 miles. The earth is not a perfect sphere but is somewhat flattened at the poles. We have provided the polar diameter since Eratosthenes' method measures the polar diameter of the earth.)

   c. Answers will vary.

2. a. Figure 4.2 in the text shows a sample simulation constructed by students who had studied right triangle trigonometry. The students affixed their short paper measuring tape to a narrow strip cut from an index card and pinned through the measuring tape to the ball, with the first pin at 0 on the measuring tape and the second pin at 4 cm. They then let their measuring tape extend from the second pin on a plane tangent to the ball.

   The light source was directly above the pin at 0, allowing the students to make a direct measure of the shadow cast by the pin at 4 cm onto the tangent segment of their measuring tape. They could then use their knowledge of right triangle trigonometry to calculate the angle at the top of the right triangle whose legs were the pin and its shadow. This angle would be the arctangent of the ratio of shadow length to pin height.

   Finally, the students used this angle in the manner of Eratosthenes to find the portion of the circumference of the ball that was represented by the 4 cm between the two pins. The calculation follows in the solution to 2(b), in which students consider the round-off error with measurements to the nearest millimeter.

   b. Suppose, for example, that the students who set up the simulation described in the solution for 2(a) measured (with all measurements precise to the nearest millimeter) the shadow as 1.6 cm, the height of the pin as 2.0 cm, and the distance between the two pins as 4.0 cm. Then the angle of the shadow is arctan(1.6/2.0) = arctan(0.8). This would yield a circumference of 360 ÷ arctan(0.8) × 4 cm, or about 37 cm.

   The students should realize that this answer is an approximation. Discarding other possible errors, such as poor technique, they might reason that because of the precision of the measurements, the true length of the shadow is between 1.55 cm and 1.65 cm. Likewise, the true pin height is between 1.95 cm and 2.05 cm. Therefore, the true ratio of the shadow length to the pin height would be between 1.55/2.05 and 1.65/1.95. This implies that the angle of the shadow (i.e., the arctangent of the ratio) would be between 37.09° and 40.24°, rounding down for the lower value and rounding up for the higher value. Continuing this line of reasoning, the students could argue that the true circumference of the ball is between 35.3 cm and 39.4 cm. They could verify their estimate by measuring the circumference directly.

3. a. Figure 4.3 in the text gives a diagram of the situation.

   b. Referring to the figure, note that because the light rays are assumed to be parallel, $m\angle B = m\angle A$ by the alternate interior angle theorem. By the exterior angle theorem, $m\angle Z + m\angle D = m\angle B = m\angle A$. Or $m\angle Z = m\angle A - m\angle D$. The students can use the lengths of the shadows to calculate $m\angle A$ and $m\angle D$, and from these measures, they can compute $m\angle Z$. Once the students determine a good estimate of the distance between the cities, they can compute the circumference as outlined in question 1.

Solutions for "Using a Distance-to-Diameter Ratio"

Moon Ratios—Part 1

1–3. In the diagram below (shown in the text as fig. 4.5), $AB \parallel DC$. Thus, $\triangle EAB \sim \triangle EDC$.

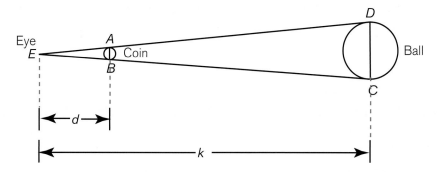

Thus, $AB/d = DC/k$. Using the tape measure, students can find AB to the nearest millimeter (e.g., a penny = 1.9 cm) and d to the nearest centimeter (e.g., 39 cm). They can then compute a value for DC/k and analyze the error.

For example, a ratio of $\frac{1.9}{39} \approx 0.049$. Since $1.85 < AB < 1.95$, and $38.5 < d < 39.5$, then $.046 < \frac{DC}{k} < .051$, rounding down for the lower bound and rounding up for the upper bound. If we use half of the interval between .046 and .051 as an estimate of the error, then .005/.049, or 10 percent, is an estimate of the relative error.

4. Students can find the visual angle by constructing a triangle similar to $\triangle EAB$ and measuring the angle. They can also use trigonometry:
$$\text{Visual angle} = 2 \times \arctan\left(\frac{0.5 \times AB}{d}\right).$$

5. a. Answers will vary.
   b. Answers will vary.
   c. Let the students discuss this. Even the direct measurements have errors. It is especially interesting when the ratio obtained by the direct measurements does not fall in the error interval computed in 3(a).

Solutions for "Using a Ratio of Time"

1. Let T_1 equal the time to full eclipse from the moment when the moon enters the earth's shadow. Note that this is the time that it takes for the moon to move a distance equal to its own diameter. Thus, if r equals the speed at which the moon is traveling, then the diameter of the moon (D_m) is equal to $r \times T_1$. Let T_2 equal the time from full eclipse to the moment when the moon completes its emergence from the earth's shadow. Note that this is the time that it takes for the moon to travel a distance approximately equal to the earth's diameter. Thus, the diameter of the earth (D_e) is equal to $r \times T_2$. This gives the following:
$$\frac{D_m}{D_e} = \frac{r \cdot T_1}{r \cdot T_2} = \frac{T_1}{T_2}.$$

2. If the diameter of the earth is 3.5 times the diameter of the moon and the polar diameter of the earth is 8400 miles, then the moon's diameter is approximately 8400/3.5, or 2400, miles. Thus, the radius of the moon is about 1200 miles. With the aid of a diagram such that in figure 4.7, students can use trigonometry to compute the distance from the earth to the moon:
$$\frac{\text{radius of moon}}{\tan\left(\frac{1}{4}°\right)} \approx 275{,}000 \text{ miles.}$$

3. a. The absolute error of this estimate is approximately 275,000 miles − 239,000 miles, or 36,000 miles.
   b. This absolute error yields a relative error of
$$\frac{275000 - 239000}{275000}, \text{ or } 13 \text{ percent.}$$

Solutions for "Figuring Out the Phases"

How Far Is the Sun?—Part 1
1–3. (Setup and process.)
4. Answers will vary. However, the range of angles for the half-moon should include 90°.
5. Yes.

Solutions for "Angling for the Distance"

How Far Is the Sun?—Part 2
1. a. The simulated moon should be very close to an exact half-moon if the simulation has been set up as in figure 4.10.
   b. Students' angle measurements will depend on their setups.
2. a. Students might agree to discard any measurements that they consider to be inaccurate because of poor technique or because the measurements are outliers from the main body of measurements. They might use the mean of the remaining measures as the angle value.

b. To estimate the error in the measurement, the students might use half of the length of the interval between the lowest and highest measurements. They might also use twice the standard deviation in the measurements, or some other justifiable estimate. Statistics texts often assume that the errors in such repeated measurements of the same phenomenon are normally distributed around 0.

3. a. Answers will vary.

   b. $\dfrac{EM}{ES} = \cos(\theta)$.

   c. Answers will vary.
   d. Students should determine an error interval for each measurement (θ and EM) and use these to determine an error interval for ES/EM.
   e. Answers will vary.

4. a. $\dfrac{EM}{ES} = \cos(87°) = 0.0523$.

   b. $ES = \dfrac{EM}{\cos(87°)} = \dfrac{239{,}000}{0.0523} \approx 4{,}570{,}000$ miles.

   c. The value of ES from 4(b)—4,570,000 miles—is only about 1/20, or approximately 5 percent, of the value used by NASA: 93,000,000 miles.

5. a. As discussed in the text, two of the sources of error that students usually mention are the measurement of the sun-earth-moon angle and the determination of exactly when the moon appears to be a half-moon.

   b. Relative error $= \dfrac{\cos(87°) - \cos(89°51'30'')}{\cos(87)} = 95$ percent.

6. a. $(1/10)° = 6$ minutes. Therefore, if Aristarchus's value for θ had been just 6 minutes greater than the reasonably close value of $89°51'30''$, he would have been working with an angle measurement of $89°57'30''$. Thus, the error in his estimate of EM/ES would have been

   $$\dfrac{\left|\cos(89°57'30'') - \cos(89°51'30'')\right|}{\cos(89°57'30'')}, \text{ or } 240 \text{ percent.}$$

   b. Aristarchus was theoretically correct in his method, but, as discussed in the text, the method was of little practical use. When an angle is close to 90°, a small error in the angle measurement makes a huge difference in the cosine ratio. Aristarchus used 87°, but the actual angle is very close to $89°51'30''$. Aristarchus thus concluded that the distance to the sun was about 19 times, (or $1/\cos(87°)$), the distance to the moon. The true ratio is closer to 400 : 1.

   c. Derived values, such as $\cos(\theta)$ in this case, can be very sensitive to small errors in the direct measurements on which they depend. When the derived value is close to 0, as in 6(a), then the relative error can be quite large, since the estimate of the error in the derived measure is divided by the derived measure, which is itself close to 0.

Solutions for "Starbucks Expansion"

1. a and b. Using the data from the Starbuck Web site, which is given on the activity sheet, we can create the following table of the number of stores that Starbucks was operating at the end of September in the years between 1987 and 2003. The data should be reliable because they are reported by the Starbucks Coffee Company, and they appear to reflect actual counts instead of numbers that have been rounded off.

Year	1987	1988	1989	1990	1991	1992	1993	1994	1995	1996	1997	1998	1999	2000	2001	2002	2003
Number of stores	17	33	55	84	116	165	272	425	676	1015	1412	1886	2135	3501	4709	5886	7225

2. a. The students' spreadsheets will look something like the data table in number 1.

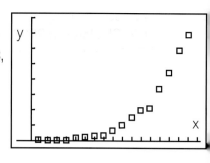

b. A scatterplot of the data on the numbers of Starbucks stores, 1987–2003, is shown at the right. *Note:* for the x-axis, instead of using the dates, we have used the number of years after the creation of Starbucks. Thus, the x-axis is scaled in one-year increments, with the first data point at $x = 1$, indicating the number of Starbucks stores after one year of operation.

3. a. The number of stores increases but not at a constant rate, since the graph is definitely not linear.
   b. If the number of stores increased at a constant rate, the graph would be linear.
   c. The rate of growth is increasing between consecutive years. The table stops with the year 2003.
   d. Answers will vary. For example, noting that between September 2002 and September 2003 approximately 1399 stores were added, it would seem to be conservative to estimate that by September 2004 there would be 7225 + 1339, or 8564, stores and that by September 2005 there would be 8564 + 1339, or 9903, stores

4. a. The graph is shown:

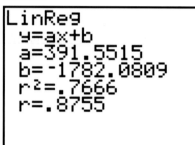 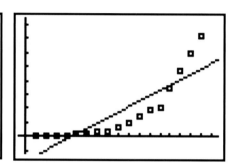

b. The line of best fit certainly does not capture the trend in the data.
c. No, the data are not linear.
d. The line of best fit would not be useful as a model for computing an estimate of the number of stores in the next several years, since its value for the year 2003 is already too low by over a thousand stores.

5. a. The following table shows the ratios of consecutive terms of the terms in the list **L2**:

Year to Previous Year	1988 to 1987	1989 to 1988	1990 to 1989	1991 to 1990	1992 to 1991	1993 to 1992	1994 to 1993	1995 to 1994	1996 to 1995	1997 to 1996	1998 to 1997	1999 to 1998	2000 to 1999	2001 to 2000	2002 to 2001	2003 to 2002
Ratios	1.94	1.67	1.53	1.38	1.42	1.65	1.56	1.59	1.50	1.39	1.34	1.13	1.64	1.35	1.25	1.23

b. These ratios are not constant, but between 1989 and 2000 they appear to hover between 1.3 and 1.6.
c. The data set does appear to be exponential between 1989 and 2001. However, it appears that the exponential growth model weakens between 2001 and 2003. The logistic model for population growth predicts this slowing in the rate of growth. The Starbucks company's long-term goal is to grow to 25,000 stores worldwide.

6. a. The graph is shown for the data given in step 1:
   b. An exponential function that is a good fit for the data follows:
   $$s(x) = A \times b^x$$
   $$s(x) = 17.78 \times 1.46^x.$$
   Note: This model suggests a 46 percent increase each year in the number of Starbucks stores.

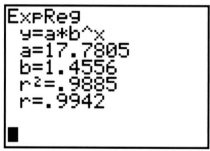 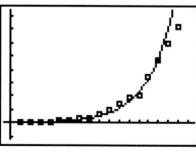

c. The growth appears to be modeled better by the exponential function.
d. Over the short term, it appears to be more accurate to use the exponential model to predict the number of stores in its chain.

a. If we assume the growth rate of 46 percent per year predicted by the exponential function in 6(b) and start with 7225 stores in 2003, then the Starbucks company would have to add 3325 stores in 2004, yielding 10,549 stores in the chain. Likewise, 2625 stores would have to be added in 2002, yielding a total of 7875 Starbucks stores.
b. By the year 2009, there would be approximately 99,980 stores in the Starbucks chain.
c. In the year 2023 there would be almost 20 million Starbucks stores. Needless to say, the exponential model will not be a suitable model for the next 20 years of the company's life. The number of stores simply cannot continue to grow exponentially at a rate of 46 percent per year.
d. The linear model makes the most sense in the long run even though it leads to questionably low predictions.

Solutions for "Golf Ball Boogie"

The points where the ball appears to "bounce" would be points on the surface. Note that the photograph doesn't capture the moment of the ball's first bounce (on the right side of the photograph). Assuming the surface is a plane, the line containing the points where the ball bounces would be on the surface.

a. Students should conclude that the ball is traveling from right to left. Students' explanations of how they knew this will vary. Many students will argue that the ball is moving from right to left since the rebound height on the leftmost bounce appears to be smaller than that of the middle bounce. Other students will notice the way in which the photograph was exposed and will conclude that images that overlap other images came later in time.
b. Answers will vary.
c. The fifth ball-image from the right edge of the photograph corresponds to the first instants in the rebound of the ball.

Students' measurements will vary depending on how they decide to measure (to the top, bottom, sides, or center of the ball-image). Students should try to be consistent in measuring from the same point in image to image. The following data were obtained by measuring from the lowest point visible on the ball to the point below it (in a vertical direction) on the surface, assuming that the bottom edge of the photograph represents the surface.

Image	0	1	2	3	4	5	6	7	8	9
Distance (cm)	0.8	2.4	3.8	5.0	5.8	6.7	7.3	7.6	7.8	7.8

Image	10	11	12	13	14	15	16	17	18
Distance (cm)	7.7	7.3	6.7	6.0	5.1	3.9	2.6	1.0	0

The 19 images represent 18 time intervals of .03 seconds each. Therefore, the ball was in the air for 5.4 seconds.

a. The circumference of an actual golf ball is about 13.5 cm. If students measure in inches, they will need to divide their measurements by .39 to convert to centimeters. The diameter of the ball is therefore about 4.3 cm.
b. The diameter of the image of the ball in the photograph is .8 cm.
c. All measurements in the table have to be multiplied by a scaling factor of about 4.3/0.8.

Using a scale factor of 4.3/0.8, students might estimate the golf ball's actual maximum distance from the surface as 40 cm.

Rounding to the nearest centimeter, students might obtain the following data:

Image	0	1	2	3	4	5	6	7	8	9
Time (sec)	0	.03	.06	.09	.12	.15	.18	.21	.24	.27
Distance (cm)	4	13	20	27	31	36	39	41	42	42

Image	10	11	12	13	14	15	16	17	18
Time (sec)	.30	.33	.36	.39	.42	.45	.48	.51	.54
Distance (cm)	41	39	37	32	27	21	14	5	0

8. Students' lists **L1** and **L2** may vary slightly, depending on the data in their tables.

9. *a* and *b*. A scatterplot created by spreadsheet software is given. The plot shows distances from the surface derived from measurements from Edgerton's photograph. In the graph, the tick marks on the *y*-axis show multiples of 5 centimeters, and the tick marks on the *x*-axis show multiples of 0.1 seconds.

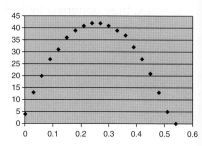

10. *a.* On the basis of the data, we can let $b = 4$ if $t = 0$, since the distance above the surface is then 4 cm. After finding the best match by varying the value of *a*, we can then add to the value for *b* to adjust the graph. The following quadratic emerges as one of many possibilities:

$$d(t) = -490t^2 + 255t + 7.$$

b. The function fits fairly well. The graph to the right shows this quadratic model determined by experimenting with values for the coefficients.

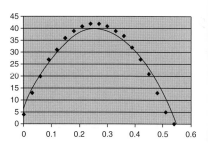

c. The units for the coefficient *a* are centimeters/second, the units for *b* are centimeters, and the units for 490 are centimeters/second^2.

Solutions for "Bouncing Ball"

1–3. (Setup and process.)

4. Sample data are given. (These data are represented in the graph to the right, which is also shown on the activity sheet.)
   a. The ball was dropped from a height of 3.93 feet.
   b. The times of the ball's first five bounces were 1.42 seconds, 2.15 seconds, 2.75 seconds, 3.23 seconds, and 3.61 seconds.
   c. The heights of the ball's first four rebounds were 2.21 feet, 1.31 feet, 0.82 feet, and 0.55 feet.

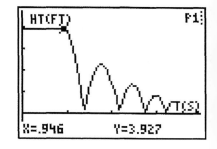

5–8. (Process.)

9. *a* and *b*. (Process.)

   c. For the sample data, the maximum point of the data plot is 2.21 feet, and the maximum point of the quadratic model is 2.22 feet. (The illustration shows the graph of selected sample data with a scatterplot showing the fit of the quadratic regression model.)

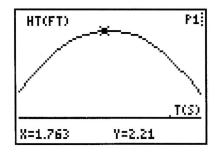
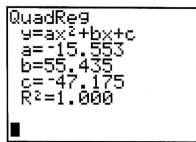
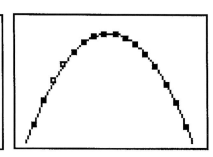

0. a. The **ZoomData** command chooses the smallest appropriate window which will show all the data.
   b. Yes, they generally look alike. When students choose different graphing windows, the calculators scale the graphs differently.
   c. The leading coefficients are generally close to −15.
   d. Students can interpret the leading coefficient in various ways. A unit analysis of the quadratic model will show that the coefficient's units are feet per second². Thus, it represents an acceleration constant. Students with an understanding of families of functions and who have explored quadratic equations will understand that the coefficient controls the curvature of the parabola.

1. Do not expect your students to give a rigorous proof using mathematical induction. If students list the first few elements of the sequence and find a pattern, then they can argue that the pattern continues. For example,

$$h_1 = \frac{2}{3}h_0, \ h_2 = \frac{2}{3}h_1 = \frac{2}{3}\left(\frac{2}{3}h_0\right) = \left(\frac{2}{3}\right)^2 h_0, \ h_3 = \frac{2}{3}h_2 = \frac{2}{3}\left(\frac{2}{3}\right)^2 h_0 = \left(\frac{2}{3}\right)^3 h_0, \ldots$$

2. a. For 3.92 feet, the initial height in the sample data, the model would predict rebound heights of 2.61 feet, 1.74 feet, 1.16 feet, and 0.78 feet. Compared with the actual rebound heights, these are too high. The elasticity and inflation of the ball affect its rebound heights.
   b. Yes, changing the fraction could improve the model.
   c. For the sample data, adjusting the rebound ratio from $\frac{2}{3}$ to $\frac{3}{5}$ gives a model that predicts heights of 2.35 feet, 1.41 feet, 0.85 feet, and 0.51 feet, much closer to the sample data.

3. a and b. If students don't take account of the shrinking times between bounces but assume that the time intervals between bounces are constant, then they are likely to think that the geometric model in step 11 implies that the ball will bounce forever. However, if they note that the time intervals are shrinking rapidly and pay attention to the sounds of the bounces, then they can argue that even though the model implies that the ball will make infinitely many bounces, the time associated with those bounces is finite.

   In fact, if we let T_n be the time that the ball is in the air in the *n*th rebound, then $\frac{1}{2}T_n$ is the time that the ball takes to fall to the ground from its rebound height h_n. However, from physics, we know that a free-falling object near the earth's surface falls a distance of $16\left(\frac{1}{2}T_n\right)^2$ feet in that amount of time. Thus,

$$h_n = 16\left(\frac{T_n}{2}\right)^2 = 4(T_n)^2.$$

   Therefore, if the rebound heights in our example really did behave according to the pattern

$$h_n = \left(\frac{3}{5}\right)^n h_0 = \left(\frac{3}{5}\right)^n \cdot 3.93 \text{ feet,}$$

   then, using the initial height from the sample data and the model from step 12(c), we would get

$$4(T_n)^2 = \left(\frac{3}{5}\right)^n \cdot 3.93,$$

   or, solving and rounding,

Thus, T_n is also a geometric sequence. Furthermore, the sum of all the rebound times,

$$\sum_{n=1}^{\infty}\left(\sqrt{\frac{3}{5}}\right)^n,$$

equals 3.5 seconds. When we add this sum to the time for the ball to drop from its initial height, our model predicts that the ball was in motion for about 4 seconds. This is comparable to the actual bounce time.

Solutions for "Most Like It Hot"

1–4. (Setup and process.) The following table shows the data obtained from one trial of the coffee-cooling experiment:

Time (minutes)	0	1	2	3	4	5	6	7	8	9
Temperature (°C)	76.7	74.6	72.4	70.3	68.3	67.0	65.2	63.9	62.5	61.1

Time (minutes)	10	11	12	13	14	15	16	17	18	19
Temperature (°C)	59.9	58.8	57.6	56.7	55.7	54.7	53.8	52.9	52.1	51.2

A scatterplot of these sample temperatures appears at the right.

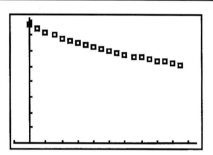

5. a. The time is measured in minutes.
   b. The temperature is measured in degrees Celsius.
   c. The temperature is decreasing as the time increases.
   d. The temperature in the sample data seems to drop quickly at first (more than 2° per minute), but later it drops more slowly (less than 1° per minute).
   e. The relationship between the time and the temperature does not appear to be linear, since the rate of decrease is not constant.
   f. No, the temperature will decrease until it reaches room temperature, but it will not go below it.
6. a. For the sample data, we recorded an ambient temperature of 20.5° C.
   b. The ambient temperature of 20.5° C is shown as a horizontal line on the following scatterplot of cooling coffee temperatures:

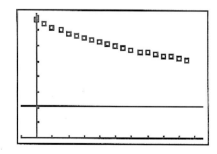

   c. An exponential function would give the best model of the temperature trends in a cup of cooling coffee.
   d. The graph shows that the temperature of the cup of coffee decreases asymptotically to a fixed value (the ambient temperature) in the manner of an exponential function.

7. a. Answers will vary. Students can choose two points in their data set to determine the line.
   b. The regression line for the sample data is $y = -1.3x + 0.74$.
   c. Using graphs, students can make visual comparisons of their linear functions with those that their calculators computed. By the criterion of least squares, the regression line is the "best fit."

8. a. Answers will vary. For the sample data, a good fit is $y = 55(0.97)^x + 20.5$.
   b. For the sample data, the exponential regression model is $y = 75(0.98)^x$.
   c. A comparison suggests that the function $y = 55(0.97)^x + 20.5$ and the exponential regression model, $y = 75(0.98)^x$, fit the sample data about equally well.

9. a. By comparing the scatterplot of the sample data with the graphs of the linear and exponential regression models, we can see that the exponential model fits better.

   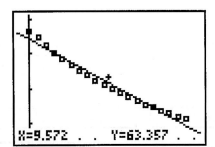 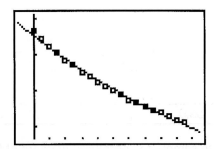

   b. The exponential model is consistent with the trend that the solution to 5(d) refers to in the differences between consecutive data points.

Navigating through Measurement in Grades 9–12   171

References

Bremigan, Elizabeth George. "Designing the Dynamic Domino Race." *Mathematics Teacher* 95 (October 2002): 502–8.

Burrill, Gail, Christine A. Franklin, Landy Godbold, and Linda J. Young. *Navigating through Data Analysis in Grades 9–12*. Principles and Standards for School Mathematics Navigations Series. Reston, Va.: National Council of Teachers of Mathematics, 2003.

Betz, William. "The Teaching of Direct Measurement in the Junior High School." In *Selected Topics in the Teaching of Mathematics*, Third Yearbook of the National Council of Teachers of Mathematics (NCTM), edited by John R. Clark and W. D. Reeve, pp. 149–94. New York: NCTM, 1928.

Day, Roger, Paul Kelley, Libby Krussel, Johnny W. Lott, and James Hirstein. *Navigating through Geometry in Grades 9–12*. Principles and Standards for School Mathematics Navigations Series. Reston, Va.: National Council of Teachers of Mathematics, 2001.

Eves, Howard. *An Introduction to the History of Mathematics*. 6th ed. Orlando, Fla.: Saunders College Publishing, 1992.

Finley, George W. "Measurement and Computation." In *Selected Topics in the Teaching of Mathematics*, Third Yearbook of the National Council of Teachers of Mathematics (NCTM), edited by John R. Clark and W. D. Reeve, pp. 141–48. New York: NCTM, 1928.

Gager, William. "Approximate Data—Terminology and Computation." In *Emerging Practices in Mathematics Education*, Twenty-second Yearbook of the National Council of Teachers of Mathematics (NCTM), edited by John R. Clark, pp. 323–38. Washington, D.C.: NCTM, 1954.

Hirshfeld, Alan W. "The Triangles of Aristarchus." *Mathematics Teacher* 97 (April 2004): 228–31.

Holliday, Berchie W., and Lauren R. Duff. "Using Graphing Calculators to Model Real-World Data." *Mathematics Teacher* 97 (May 2004): 328–42.

House, Peggy A., and Roger P. Day, eds. *Mission Mathematics II: Grades 9–12*. Reston, Va.: National Council of Teachers of Mathematics, 2005.

Kerr, Donald R., Jr., and Frank K. Lester, Jr. "An Error Analysis Model for Measurement." In *Measurement in School Mathematics*, 1976 Yearbook of the National Council of Teachers of Mathematics (NCTM), edited by Doyal Nelson, pp. 105–22. Reston, Va.: NCTM, 1976.

Moyer, Patricia S., and Wei Shen Hsia. "The Archaeological Dig Site: Using Geometry to Reconstruct the Past." *Mathematics Teacher* 94 (March 2001): 193–99; 206–7.

National Council of Teachers of Mathematics (NCTM). *Principles and Standards for School Mathematics*. Reston, Va.: NCTM, 2000.

National Research Council. *National Science Education Standards*. Washington, D.C.: National Academy Press, 1996.

Neugebauer, Otto. *A History of Ancient Mathematical Astronomy*. Part 2. New York: Springer-Verlag, 1975.

O'Connor, James J. "Forever May Only Be a Few Seconds." *Mathematics Teacher* 92 (April 1999): 300–301.

Payne, Joseph N., and Robert C. Seber. "Measurement and Approximation." In *The Growth of Mathematical Ideas: Grades K–12*, Twenty-fourth Yearbook of the National Council of Teachers of Mathematics (NCTM), edited by Phillip S. Jones et al., pp. 182–228. Washington, D.C.: NCTM, 1959.

Shuster, C. N. "The Use of Measuring Instruments in Teaching Mathematics." In *Selected Topics in the Teaching of Mathematics*, Third Yearbook of the National Council of Teachers of Mathematics (NCTM), edited by John R. Clark and W. D. Reeve, pp. 195–222. New York: NCTM, 1928.

Shuster, Carl. "Working with Approximate Data." In *Emerging Practices in Mathematics Education*, Twenty-second Yearbook of the National Council of Teachers of Mathematics (NCTM), edited by John R. Clark, pp. 310–23. Washington, D.C.: NCTM, 1954.

Stephens, Gregory P. "Teaching the Logistic Function in High School." *Mathematics Teacher* 95 (April 2002): 286–94.

Suggested Reading

Albrecht, Masha. "The Volume of a Pyramid: Low-Tech and High-Tech Approaches." *Mathematics Teacher* 94 (January 2001): 58–64.

Boyer, Carl B. *A History of Mathematics*. Princeton, N.J.: Princeton University Press, 1985.

Dauben, Joseph W. *Historical Notes: Mathematics through the Ages*. Lexington, Mass.: Consortium for Mathematics and Its Applications (COMAP), 1992.

Ferguson, Kitty. *Measuring the Universe: Our Historic Quest to Chart the Horizons of Space and Time*. New York: Walker Publishing Company, 1999.

Glanville, James O. *General Chemistry for Engineers: Second Preliminary Edition*. Upper Saddle River, N.J.: Prentice Hall, 2002.

Peitgen, Heinz-Otto, Hartmut Jürgens, Dietmar Saupe, Evan Maletsky, and Terry Perciante. *Fractals for the Classroom: Strategic Activities*. Vol. 3. New York and Reston, Va.: Springer-Verlag and National Council for Teachers of Mathematics, 1999.

Perdew, Patrick R. "Sports and Distance-Rate-Time." *Mathematics Teacher* 95 (March 2002): 192–99.

Sobel, Max A., and Evan M. Maletsky. "Iteration Activities and Fractal Patterns." Chap. 8 in *Teaching Mathematics: A Sourcebook of Aids, Activities, and Strategies*. Needham Heights, Mass.: Allyn & Bacon, 1999.

Taylor, Barry N., and Chris E. Kuyatt. *Guidelines for Evaluating and Expressing the Uncertainty of NIST Measurement Results*. NIST Technical Note 1297. Gaithersburg, Md.: National Institute of Standards and Technology, 1994. Available in pdf format at http://physics.nist.gov/Document/tn1297.pdf.